The Connective K-Theory of Finite Groups

MEMOIRS
of the
American Mathematical Society

Number 785

The Connective K-Theory
of Finite Groups

R. R. Bruner
J. P. C. Greenlees

American Mathematical Society
Providence, Rhode Island

2000 *Mathematics Subject Classification.* Primary 19L41, 19L47, 19L64, 55N15; Secondary 20J06, 55N22, 55N91, 55T15, 55U20, 55U30.

Library of Congress Cataloging-in-Publication Data

Bruner, R. R. (Robert Ray), 1950–
  The connective K-theory of finite groups / R. R. Bruner, J. P. C. Greenlees.
    p. cm. — (Memoirs of the American Mathematical Society, ISSN 0065-9266 ; no. 785)
  "Volume 165, number 785 (second of 4 numbers)."
  Includes bibliographical references and indexes.
  ISBN 0-8218-3366-9 (alk. paper)
  1. K-theory. 2. Finite groups. I. Greenlees, J. P. C. (John Patrick Campbell), 1959– II. Title. III. Series.

QA3.A57 no. 785
[QA612.33]
510s—dc21
[514′.23]   2003051902

Memoirs of the American Mathematical Society

This journal is devoted entirely to research in pure and applied mathematics.

Subscription information. The 2003 subscription begins with volume 161 and consists of six mailings, each containing one or more numbers. Subscription prices for 2003 are $555 list, $444 institutional member. A late charge of 10% of the subscription price will be imposed on orders received from nonmembers after January 1 of the subscription year. Subscribers outside the United States and India must pay a postage surcharge of $31; subscribers in India must pay a postage surcharge of $43. Expedited delivery to destinations in North America $35; elsewhere $130. Each number may be ordered separately; *please specify number* when ordering an individual number. For prices and titles of recently released numbers, see the New Publications sections of the *Notices of the American Mathematical Society*.

Back number information. For back issues see the *AMS Catalog of Publications*.

Subscriptions and orders should be addressed to the American Mathematical Society, P. O. Box 845904, Boston, MA 02284-5904, USA. *All orders must be accompanied by payment.* Other correspondence should be addressed to 201 Charles Street, Providence, RI 02904-2294, USA.

Copying and reprinting. Individual readers of this publication, and nonprofit libraries acting for them, are permitted to make fair use of the material, such as to copy a chapter for use in teaching or research. Permission is granted to quote brief passages from this publication in reviews, provided the customary acknowledgment of the source is given.

Republication, systematic copying, or multiple reproduction of any material in this publication is permitted only under license from the American Mathematical Society. Requests for such permission should be addressed to the Acquisitions Department, American Mathematical Society, 201 Charles Street, Providence, Rhode Island 02904-2294, USA. Requests can also be made by e-mail to `reprint-permission@ams.org`.

Memoirs of the American Mathematical Society is published bimonthly (each volume consisting usually of more than one number) by the American Mathematical Society at 201 Charles Street, Providence, RI 02904-2294, USA. Periodicals postage paid at Providence, RI. Postmaster: Send address changes to Memoirs, American Mathematical Society, 201 Charles Street, Providence, RI 02904-2294, USA.

© 2003 by the American Mathematical Society. All rights reserved.
This publication is indexed in *Science Citation Index*®, *SciSearch*®, *Research Alert*®, *CompuMath Citation Index*®, *Current Contents*®/*Physical, Chemical & Earth Sciences*.
Printed in the United States of America.

∞ The paper used in this book is acid-free and falls within the guidelines
established to ensure permanence and durability.
Visit the AMS home page at `http://www.ams.org/`

10 9 8 7 6 5 4 3 2 1   08 07 06 05 04 03

Contents

Chapter 0. Introduction. ... 1
  0.1. Motivation. ... 1
  0.2. Highlights of Chapter 1. ... 2
  0.3. Highlights of Chapter 2. ... 3
  0.4. Highlights of Chapter 3. ... 4
  0.5. Highlights of Chapter 4. ... 5
  0.6. Reading guide. ... 6
  0.7. Acknowledgements. ... 6

Chapter 1. General properties of the ku-cohomology of finite groups. ... 7
  1.1. Varieties for connective K-theory. ... 7
  1.2. Implications for minimal primes. ... 11
  1.3. Euler classes and Chern classes. ... 14
  1.4. Bockstein spectral sequences. ... 18
  1.5. The Künneth theorem. ... 21

Chapter 2. Examples of ku-cohomology of finite groups. ... 27
  2.1. The technique. ... 28
  2.2. Cyclic groups. ... 32
  2.3. Nonabelian groups of order pq. ... 36
  2.4. Quaternion groups. ... 43
  2.5. Dihedral groups. ... 51
  2.6. The alternating group of degree 4. ... 59

Chapter 3. The ku-homology of finite groups. ... 63
  3.1. General behaviour of $ku_*(BG)$. ... 63
  3.2. The universal coefficient theorem. ... 66
  3.3. Local cohomology and duality. ... 68
  3.4. The ku-homology of cyclic and quaternion groups. ... 69
  3.5. The ku-homology of BD_8. ... 71
  3.6. Tate cohomology. ... 76

Chapter 4. The ku-homology and ku-cohomology of elementary abelian groups. ... 79
  4.1. Description of results. ... 79
  4.2. The ku-cohomology of elementary abelian groups. ... 81
  4.3. What local cohomology ought to look like. ... 87
  4.4. The local cohomology of Q. ... 88
  4.5. The 2-adic filtration of the local cohomology of Q. ... 93
  4.6. A free resolution of T. ... 94

4.7.	The local cohomology of T.	99
4.8.	Hilbert series.	102
4.9.	The quotient P/T_2.	103
4.10.	The local cohomology of R.	104
4.11.	The ku-homology of BV.	105
4.12.	Duality for the cohomology of elementary abelian groups.	109
4.13.	Tate cohomology of elementary abelian groups.	111
Appendix A.	Conventions.	115
A.1.	General conventions.	115
A.2.	Adams spectral sequence conventions.	115
Appendix B.	Indices.	117
B.1.	Index of calculations.	117
B.2.	Index of symbols.	117
B.3.	Index of notation.	118
B.4.	Index of terminology.	122
Appendix.	Bibliography	125

Abstract

This paper is devoted to the connective K homology and cohomology of finite groups G. We attempt to give a systematic account from several points of view.

In Chapter 1, following Quillen [50, 51], we use the methods of algebraic geometry to study the ring $ku^*(BG)$ where ku denotes connective complex K-theory. We describe the variety in terms of the category of abelian p-subgroups of G for primes p dividing the group order. As may be expected, the variety is obtained by splicing that of periodic complex K-theory and that of integral ordinary homology, however the way these parts fit together is of interest in itself. The main technical obstacle is that the Künneth spectral sequence does not collapse, so we have to show that it collapses up to isomorphism of varieties.

In Chapter 2 we give several families of new complete and explicit calculations of the ring $ku^*(BG)$. This illustrates the general results of Chapter 1 and their limitations.

In Chapter 3 we consider the associated homology $ku_*(BG)$. We identify this as a module over $ku^*(BG)$ by using the local cohomology spectral sequence. This gives new specific calculations, but also illuminating structural information, including remarkable duality properties.

Finally, in Chapter 4 we make a particular study of elementary abelian groups V. Despite the group-theoretic simplicity of V, the detailed calculation of $ku^*(BV)$ and $ku_*(BV)$ exposes a very intricate structure, and gives a striking illustration of our methods. Unlike earlier work, our description is natural for the action of $GL(V)$.

Received by the editor January 1, 2002.

1991 *Mathematics Subject Classification.* Primary: 19L41, 19L47, 19L64, 55N15; Secondary: 20J06, 55N22, 55N91, 55T15, 55U20, 55U25, 55U30.

Key words and phrases. connective K-theory, finite group, cohomology, representation theory, Chern classes, Euler classes, cohomological varieties, Gorenstein, local cohomology, dihedral, quaternion, elementary abelian.

The first author is grateful to the EPSRC, the Centre de Recerca Matemàtica, and the Japan-US Mathematics Institute, and the second author is grateful to the Nuffield Foundation and the NSF for support during work on this paper.

CHAPTER 0

Introduction.

0.1. Motivation.

This paper is about the connective complex K theory $ku^*(BG)$ and $ku_*(BG)$ of finite groups G. The first author and others [3, 4, 5, 6, 8, 35, 48] have made many additive calculations of $ku^*(BG)$ and $ku_*(BG)$ for particular finite groups G. The purpose of the present paper is to give a more systematic account.

More precisely, $ku^*(X)$ is the cohomology represented by the connective cover of the spectrum K representing Atiyah-Hirzebruch periodic complex K theory. Their values on a point are

$$K_* = \mathbb{Z}[v, v^{-1}] \text{ and } ku_* = \mathbb{Z}[v],$$

where v is the Bott periodicity element in degree 2.

Connective K theory is relatively easy to calculate, and it has been used to great effect as a powerful and practical invariant by homotopy theorists. However, it is not well understood from a theoretical point of view. Although it can be constructed by infinite loop space theory, and there are ad hoc interpretations of its values in terms of vector bundles trivial over certain skeleta [54], these fall short of a satisfactory answer (for instance because they fail to suggest a well behaved equivariant analogue). Similarly, ku is also a complex oriented theory, and is the geometric realization of the representing ring for multiplicative formal groups if we allow a non-invertible parameter. This does appear to generalize to the equivariant case [27] but, since ku_* is not Landweber exact, it does not give a definition.

Our response to this state of affairs is to exploit the calculability. In practical terms this extends the applicability of the theory, and in the process we are guided in our search for geometric understanding.

Our results fall into four types, corresponding to the four chapters.

(1) General results about the cohomology rings $ku^*(BG)$, describing its variety after Quillen. In the course of this we exploit a number of interesting general properties, such as the behaviour of the Bockstein spectral sequence and the fact that the Künneth theorem holds up to nilpotents for products of cyclic groups. The use of Euler classes of representations is fundamental.
(2) Explicit calculations of cohomology rings for low rank groups. The input for this is the known group cohomology ring (processed via the Adams spectral sequence) and the complex character ring.
(3) General results about the homology modules $ku_*(BG)$, and the curious duality phenomena which appear. The connections with the cohomology

ring via the local cohomology theorem are highlighted. Phenomena are illustrated by specific examples.

(4) Finally we turn to elementary abelian groups. We calculate the cohomology ring and the homology module. These turn out to be remarkably complicated, and the detailed geometry is quite intricate. Nonetheless we are able to give a complete analysis which is a good guide for the complexities arising in more general cases.

There are several reasons for interest in results of this type. Our results involve the introduction of a 'character ring' together with a character map $ku^*(BG) \longrightarrow \hat{Ch}_{ku}(G)$. The point is that the character ring is reasonably calculable, and the kernel of the character map consists of nilpotents so we can obtain useful new information on the ring structure of $ku^*(BG)$.

Secondly, there are equivariant versions of connective K-theory. For any reasonable version, a completion theorem holds in the form

$$ku^*(BG) = (ku_G^*)_I^\wedge$$

for a suitable ideal I. One example is May's MU-induction [45] of non-equivariant ku [32], but this equivariant version is much too big to be entirely satisfactory. As a result of the present work, the second author has constructed a better version for groups of prime order in [25], and for general compact Lie groups G in [28]. In any case, the results of the present paper give information about the completion of the coefficient ring ku_G^* for any of these theories. Our calculations show that a number of naive expectations are unreasonable, and also provide calculations of the coefficient rings of the equivariant theory of [28].

Thirdly, the case of connective K-theory is of interest as an introduction to some of the methods of [33], and a step on the way to extending them to the general case of $BP\langle n \rangle$. N.P.Strickland [56] has calculations in $BP\langle n \rangle^* BV$ when V is elementary abelian of rank less than $n + 2$, suggesting further progress may be possible.

0.2. Highlights of Chapter 1.

In the first chapter we prove a number of general results giving a description of the coarser features of $ku^*(BG)$ as a ring in terms of the group theory of G. Quillen [50, 51] has given a descent argument which shows that the variety of the ring $ku^*(BG)$ can be calculated from that of $ku^*(BA)$ for abelian subgroups A using the category of abelian subgroups of G. This shows that most of our general purpose can be achieved if we can calculate $ku^*(BA)$ for abelian groups A and describe it in a functorial way. This is easy if A is cyclic. For other cohomology theories that have been analyzed, the general case then follows by the Künneth theorem, but the Künneth spectral sequence for ku does not generally collapse to a tensor product decomposition. Nonetheless, by a fairly elaborate argument using special features of connective K-theory, we can show that we do have a tensor product decomposition up to varieties, and this allows us to give useful general results.

As might be expected, the answer is that the variety is a mixture of that of ordinary cohomology at various integer primes and that of K theory. Quillen proved that the mod p cohomology ring has dimension equal to the p-rank of G, whilst periodic K-theory is one dimensional. It follows that $ku^*(BG)$ has dimension equal to the rank of G if G is non-trivial, and the variety is formed by sticking the variety of $K^*(BG)$ (with components corresponding to conjugacy classes of cyclic subgroups of G) to those of $H^*(BG; \mathbb{F}_p)$ (with components corresponding to conjugacy classes of maximal elementary abelian p-subgroups of G). Considering the degenerations as v or p becomes zero, it is not hard to identify the components of the variety of $ku^*(BG)$ or equivalently the minimal primes of the ring.

The ingredients in this analysis are interesting. First, we use Euler classes and Chern classes of representations. It is very valuable to have a good supply of classes under tight control, and we discuss their behaviour under a number of natural constructions such as tensor products and change of cohomology theory.

In view of the coefficient ring $ku_* = \mathbb{Z}[v]$ it is natural to study $ku^*(X)$ according to its p- and v-torsion, and in both cases the Bockstein spectral sequence gives a structure for understanding this torsion. For integral ordinary homology it is not hard to use the Bockstein spectral sequence to show that the Künneth theorem holds up to varieties; nonetheless, this is not a formality. The analogous result for connective K theory involves more careful analysis, and comparison with periodic K theory. Furthermore we need to compare a number of filtrations that occur naturally and show they coincide in favourable cases.

0.3. Highlights of Chapter 2.

Next we give exact calculations of the cohomology $ku^*(BG)$ as a ring for a number of groups of small rank. The only examples previously known were the cyclic groups, where the result is clear from complex orientability and the Gysin sequence. We are obliged to use more elaborate methods.

The additive structure largely follows from the Adams spectral sequence. Because the mod p cohomology of ku is free over $\mathcal{A}//E(Q_0, Q_1)$, where $\mathcal{A}$ is the mod p Steenrod algebra and $E(Q_0, Q_1)$ is the exterior algebra on Q_0 and Q_1, the E_2-term of the Adams spectral sequence for $ku^*(BG)$ only requires the ordinary mod p cohomology groups as modules over this exterior algebra. The higher Bocksteins, the action of the rest of $\mathcal{A}$, and a simple fact about the action of the representation ring on $ku^*(BG)$ (Lemma 2.1.1) determine the differentials, and this gives $ku^*(BG)$ up to extensions. The multiplicative relations then follow by this lemma and by comparison with ordinary cohomology and the representation ring. This step is most effective for periodic groups, and for this reason we restrict attention to low rank groups (which are also those with low dimensional cohomology ring). Amongst rank 1 groups we calculate the cohomology of the cyclic and generalized quaternion groups, and the non-abelian groups of order pq. We then proceed to calculate the cohomology ring of the dihedral groups and the alternating group A_4.

For rank 1 groups, ku^*BG is a subring of K^*BG, so that once we have identified the generators of ku^*BG using the Adams spectral sequence, we can use K^*BG and $R(G)$ to determine the relations. Even so, this produces rather complicated relations when the order is large. The appearance of the Chebyshev polynomials in the relations satisfied by the quaternion and dihedral groups is somewhat surprising (2.5.3), and the number theory involved in the non-abelian groups of order pq is tantalizing.

Among higher rank groups, the calculation is complicated by the presence of (p,v)-torsion. This makes direct use of the Adams spectral sequence less helpful on the face of it. However, in each of the cases we consider, the Adams spectral sequence does show that there is no (v)-torsion in positive Adams filtration, and hence that cohomology together with representation theory will completely determine the multiplicative structure. It is interesting to note that some of the p-torsion in integral cohomology comes from (p,v)-torsion in ku^*BG, whilst other p-torsion in cohomology reflects v-divisibility of non-(p,v)-torsion in ku^*BG. The quaternion and dihedral groups in particular, show the relation between the order of torsion in cohomology and the v-filtration of the representation ring (Remark 2.5.10). All of the (2)-torsion in $ku^*BD_{2^n}$ is of order exactly 2, is annihilated by v, and is independent of n. The remaining integral cohomology (in the image of reduction from ku) is of order 2^{n-1}, reflecting divisibility by v. Similarly, $ku^*BQ_{2^n}$ has no (2) or (v)-torsion at all, whereas $H\mathbb{Z}^*BQ_{2^n}$ has torsion of orders 2 and 2^n, reflecting v-divisibility of corresponding classes in $ku^*BQ_{2^n}$.

Finally, as in ordinary cohomology, $ku^*(BA_4)$ is not generated by Chern classes. In addition, the Euler class of the simple representation of dimension 3 is non-trivial in $ku^*(BA_4)$ but vanishes in periodic K-theory. An alternative approach to $ku^*(BA_4)$ is to note that the 2-local part of the $ku^*(BA_4)$ is exactly the C_3 invariants in ku^*BV_4, where V_4 is the Klein 4-group.

One interesting general pattern that emerges may be summarized by considering the short exact sequence

$$0 \longrightarrow T \longrightarrow ku^*(BG) \longrightarrow Q \longrightarrow 0$$

of $ku^*(BG)$-modules where T is the v-power torsion. Thus Q is the image of $ku^*(BG)$ in $K^*(BG)$, and in all our examples except A_4 it is the $K^0(BG)$ subalgebra of $K^*(BG)$ generated by $1, v$ and the Chern classes of representations. Although T is defined as the v-power torsion, in our examples it turns out to be the (p,v)-power torsion when G is a p-group. In many cases it is annihilated by the exponent of group without the need for higher powers.

0.4. Highlights of Chapter 3.

Next we consider homology $ku_*(BG)$, and especially how the cohomology ring $ku^*(BG)$ affects it. It is generally considered that the homology of a group is more complicated than its cohomology since it involves various forms of higher torsion. However one of the lessons of our approach is that the right commutative algebra shows that the two contain the same information up to duality.

Earlier results give just additive information, and our emphasis is on structural properties by use of the local cohomology spectral sequence, which allows us to deduce $ku_*(BG)$ as a module over $ku^*(BG)$ from a knowledge of the ring $ku^*(BG)$ and the Euler classes. We emphasize that although this has purely practical advantages in the ease of calculating certain additive extensions, the main attraction is the structural and geometric information not accessible through the Adams spectral sequence. To explain, we let $I = \ker(ku^*(BG) \longrightarrow ku^*)$ denote the augmentation ideal. Local cohomology $H_I^*(M)$ is a functor on $ku^*(BG)$ modules M, and calculates right derived functors of the I-power torsion functor

$$\Gamma_I(M) = \{m \in M \mid I^s m = 0 \text{ for } s >> 0\}$$

in the sense that

$$H_I^*(M) = R^*\Gamma_I(M).$$

The local cohomology modules $H^i_I(M)$ vanish for $i > \dim(ku^*(BG))$, and so the local cohomology spectral sequence
$$H^*_I(ku^*(BG)) \Rightarrow ku_*(BG)$$
is a finite spectral sequence. The E_2-term is calculable and the whole spectral sequence is natural in G: we use both these facts to great effect.

In fact the local cohomology spectral sequence is a manifestation of a remarkable duality property of the ring $ku^*(BG)$. For ordinary mod p cohomology the corresponding duality implies, for example, that a Cohen-Macaulay cohomology ring is automatically Gorenstein. Since ku^* is more complicated than $H\mathbb{F}_p^*$, the statement is more complicated for connective K theory, but the phenomenon is nonetheless very striking. This is reflected again in Tate cohomology. As with ordinary cohomology, the advantages of Tate cohomology are most striking in the rank 1 case. More precisely, one may combine the universal coefficient spectral sequence
$$\mathrm{Ext}^{*,*}_{ku_*}(ku_*(BG), ku_*) \Rightarrow ku^*(BG),$$
with the local cohomology theorem to make a statement of the form
$$(\mathrm{RD}) \circ (\mathrm{R}\Gamma_I)(ku^*(BG)) \text{ '='} ku^*(BG).$$

Here $D(\cdot) = \mathrm{Hom}_{ku_*}(\cdot, ku_*)$ denotes duality, R denotes the right derived functor in some derived category, and the precise statement takes place in a category of strict modules over the S^0-algebra $F(BG_+, ku)$, or its equivariant counterpart $F(EG_+, ku)$. This states that the commutative equivariant S^0-algebra $F(EG_+, ku)$ is 'homotopy Gorenstein' in the sense of [15], and this has structural implications for its coefficient ring $ku^*(BG)$. We give this heuristic discussion some substance by showing what it means in practice for our particular examples from the previous chapter. It turns out that under the local cohomology theorem, the Universal coefficient short exact sequence calculating $\widetilde{ku}^*(BG)$ from $\widetilde{ku}_*(BG)$ corresponds exactly to the short exact sequence $0 \longrightarrow T \longrightarrow ku^*(BG) \longrightarrow Q \longrightarrow 0$ mentioned in the summary of Chapter 3.

0.5. Highlights of Chapter 4.

Finally we discuss a considerably more complicated case. Although there is very little to the group theory of elementary abelian groups, it interacts in a quite intricate way with the coefficient ring and provides a test for the effectiveness of our methods. The connective case is more complicated because of the failure of the exact Künneth theorem. Geometrically speaking, this means that the basic building block is no longer affine space. It is intriguing to see the commutative algebra that this gives rise to, and it is chastening to see the complexity that so small a perturbation causes.

Again we consider the extension
$$0 \longrightarrow T \longrightarrow ku^*(BV) \longrightarrow Q \longrightarrow 0$$
of $ku^*(BV)$-modules, where T is the v-power torsion, which in this case is the same as the ideal of elements annihilated by (p, v). Then Q has no p or v-torsion. Indeed, Q is the image of $ku^*(BV)$ in periodic K-theory: it is the Rees ring, which is to say that it is the $K^0(BG)$ subalgebra of $K^*(BG)$ generated by $1, v$ and the Euler classes $y_1, y_2, \ldots, y_r$ of any r generating one dimensional representations. Additively, Q is the sum of the coefficient ring $\mathbb{Z}[v]$ and a module which is free of

rank $|V|-1$ over $\mathbb{Z}_p^\wedge$ in each even degree. Finally, if $p=2$, the module T is the ideal of $\mathbb{F}_2[x_1, x_2, \ldots, x_r]$ generated by the elements $q_S = Q_1 Q_0 (\prod_{s \in S} x_s)$ with S a subset of $\{1, 2, \ldots, r\}$, and the action of the Euler classes on T is determined by the fact y_i acts as x_i^2. This determines the ring structure of $ku^*(BV)$. However, the commutative algebra of $ku^*(BV)$ is rather complicated. It turns out that there is a direct sum decomposition $T = T_2 \oplus T_3 \oplus \cdots \oplus T_r$, and T_i is of dimension r and depth i. (Notice that this decomposition and the following discussion is intrinsic and natural in V). Remarkably, the local cohomology of T_i is concentrated in degrees i and r. More startling still is the duality: the subquotients of $H_I^1(Q)$ under the 2-adic filtration are the modules $H_I^i(T_i)$, and the differentials in the local cohomology spectral sequence give the isomorphism (4.11.5). Furthermore the top local cohomology groups pair up $H_I^r(T_i)^\vee = \Sigma^{-r+4} T_{r-i+2}$ for $i = 2, 3, \ldots, r-1$ (the exceptional behaviour of T_r is exactly what is required to lead to a clean final duality statement): there is a natural exact sequence

$$0 \longrightarrow \Sigma^{-4} T^\vee \longrightarrow ku_*(BV) \longrightarrow \mathbb{Z}[v] \oplus \Sigma^{-1}(2^{r-1} H_I^1(Q)) \longrightarrow 0.$$

Finding duality in a rank 1 group might seem like little more than a coincidence, but in these rank r groups the exceptional behaviour necessary is quite breathtaking.

0.6. Reading guide.

Although the chapters are largely independent, certain sections are applied in later chapters. Thus all readers will need to read Section 1.1 for its basic facts and Section 1.3 for its discussion of Euler and Chern classes. Readers interested in the homology of explicit examples in Chapter 3 will need to read the corresponding sections in Chapter 2. Readers of Chapters 4 will want to read Section 2.1 to explain how we calculate cohomology and hence introduce Section 4.2, and those interested in homology and duality will also need to read most of Chapter 3.

We direct readers to Appendix A where we summarize various conventions. In Appendix B there are various indices that should assist selective readers.

0.7. Acknowledgements.

Thanks are due to Peter Malcolmson for always being willing to listen, and often having useful suggestions, to Lowell Hansen for the observation that the interesting polynomials we had must be conjugates of the Chebyshev polynomials, and to Hugh Montgomery for pointing us to the class number formula.

Special thanks are due to an exceptionally diligent referee who made a number of useful suggestions.

CHAPTER 1

General properties of the ku-cohomology of finite groups.

This chapter discusses generalities, but many of them will be used in our specific calculations in later chapters. The main theme is to investigate what Quillen's method proves about the variety of $ku^*(BG)$. Since ku is complex oriented, the structure of the argument is exactly Quillen's, and this is outlined in Section 1.1. The implications for minimal primes of $ku^*(BG)$ are described in Section 1.2, and compared with analogous results for periodic K theory and ordinary cohomology. Euler classes provide a basic ingredient and point of comparison between theories, and we devote Section 1.3 to them. The principal difficulty in implementing Quillen's argument is that the Künneth theorem does not hold exactly, so that the cohomology of abelian groups is not the tensor product of the cohomology of cyclic groups. Using an analysis of the Bockstein spectral sequence in Section 1.4 we are able in Section 1.5 to show that the Künneth theorem does hold up to nilpotents in relevant cases.

1.1. Varieties for connective K-theory.

We shall be concerned with ku^*BG for finite groups G. Here, $ku^*(\cdot)$ is the unreduced connective complex K-theory with coefficient ring $ku_* = \mathbb{Z}[v]$ where $v \in ku^{-2} = ku_2$. Because we need to discuss both homology and cohomology, it is essential for clarity that we consistently refer to cohomological degrees (i.e., 'upper degrees' in the sense of Cartan-Eilenberg) as codegrees. Degrees can be expressed as codegrees in the usual way, by $M^k = M_{-k}$: for example, v has degree 2 and codegree -2.

Before going further we should record the relationship between ku and the more familiar theories, ordinary integral cohomology $H\mathbb{Z}$, and periodic K-theory K. Indeed, there is a cofibre sequence

$$\Sigma^2 ku \xrightarrow{v} ku \longrightarrow H\mathbb{Z},$$

and there is an equivalence

$$K \simeq ku[1/v].$$

The full relationship is described by a Bockstein spectral sequence which we shall have occasion to consider below, but for the present we shall be satisfied with two remarks. First, note that the cofibre sequence gives a short exact sequence

$$0 \longrightarrow ku^*(X)/(v) \longrightarrow H\mathbb{Z}^*(X) \longrightarrow \mathrm{ann}(v, ku^*(\Sigma^{-3}X)) \longrightarrow 0.$$

LEMMA 1.1.1. *For any space X we have $(ku^*X)[1/v] \xrightarrow{\cong} K^*(X)$.*

Proof: We prove the result for any bounded below spectrum X. Indeed, the natural map $ku \longrightarrow ku[1/v] \simeq K$ gives a natural transformation
$$ku^*(X)[1/v] \longrightarrow K^*(X),$$
since v is invertible in the codomain. This is tautologically an isomorphism when X is a sphere. Both domain and codomain are exact functors of X, and $K^*(X)$ satisfies the wedge axiom. It therefore suffices to show that $ku^*(X)[1/v]$ satisfies the wedge axiom for bounded below wedges in the sense that when X_i is n-connected for all i, the natural map
$$ku^*(\bigvee_i X_i)[1/v] \xrightarrow{\cong} \prod_i (ku^*(X_i)[1/v])$$
is an isomorphism.

This holds since the relevant limits are achieved in each degree. More precisely, inverting v in $ku^s Y$ involves passing to limits over the sequence
$$ku^s(Y) \longrightarrow ku^{s-2}(Y) \longrightarrow ku^{s-4}(Y) \longrightarrow ku^{s-6}(Y) \longrightarrow \cdots.$$
If Y is n-connected, the maps are all isomorphic once $s - 2k \leq n$. $\square$

We now briefly summarize Quillen's methods [50, 51], explaining how they apply to ku. First note that ku is complex orientable. Next, the usual argument with the fibration $U(n)/G \longrightarrow BG \longrightarrow BU(n)$ associated to a faithful representation of G in $U(n)$ shows that $ku^*(BG)$ is Noetherian, and similarly one sees that if Z is any finite G-complex, the cohomology $ku^*(EG \times_G Z)$ is a finitely generated module over it. Accordingly, we may apply Quillen's descent argument to deduce that the restriction map
$$ku^* BG \longrightarrow \varprojlim_A ku^* BA =: \hat{Ch}(G)$$
is a V-isomorphism (in the sense that it induces an isomorphism of varieties), and that the variety of the inverse limit is the direct limit of the varieties of the terms $ku^*(BA)$.

More precisely, we have to bear in mind that $ku^*(BG)$ is endowed with the skeletal topology, so it is appropriate to consider formal schemes. For any ring R with linear topology, we have an associated formal scheme $\mathrm{spf}(R)$, defined as a functor from topological rings k to sets by
$$\mathrm{spf}(R)(k) = \mathrm{Hom}_{cts}(R, k).$$
The example to hand is
$$\mathfrak{X}(G) = \mathrm{spf}(ku^*(BG)),$$
which is a formal scheme over $\mathfrak{X} = \mathrm{spec}(ku^*)$. The underlying variety is obtained by restricting the functor to indiscrete algebraically closed fields, and since all our results are at the level of varieties the reader may forget the topology on $ku^* BG$ for our purposes. There is always a map $\varinjlim_i \mathrm{spf}(R_i) \longrightarrow \mathrm{spf}(\varprojlim_i R_i)$, and Quillen has shown that if we restrict to a finite diagram of algebras R_i, finite over the inverse limit, it is an isomorphism of varieties. Thus Quillen's theorem states that the natural map
$$\varinjlim_A \mathfrak{X}(A) \xrightarrow{V\cong} \mathfrak{X}(G)$$

of schemes over ku^* is an isomorphism at the level of varieties. To analyze $\mathfrak{X}(G)$ as a variety it is enough to understand $\mathfrak{X}(A)$ for abelian groups A in a functorial way.

Consider the special case $R = ku^*BS^1 \cong ku^*[[x]]$, which is free as a topological ring on a single topologically nilpotent generator x. Thus
$$\mathrm{spf}(ku^*[[x]])(k) = \mathrm{nil}(k),$$
where $\mathrm{nil}(k)$ is the set of topologically nilpotent elements of k. The map $BS^1 \times BS^1 \longrightarrow BS^1$ classifying tensor product of line bundles makes $ku^*[[x]]$ into a Hopf algebra with coproduct
$$x \longmapsto 1 \otimes x + x \otimes 1 - vx \otimes x,$$
and thus $\mathrm{nil}(k)$ is naturally a group under $x \odot y = x + y - vxy$. We write $[n](x)$ for the n-fold $\odot$ sum of x with itself, so that
$$[n](x) = (1 - (1-vx)^n)/v.$$

Let k^{A^*} denote the ring of functions from A^* to k under pointwise multiplication. We may give it the structure of a Hopf algebra with coproduct from the product in A^*:
$$\psi(f)(\beta \otimes \gamma) = f(\beta\gamma).$$

THEOREM 1.1.2. *For abelian groups A there is a map*
$$\mathrm{spf}(ku^*BA)(k) \xrightarrow{V \cong} \mathrm{Hopf}_{cts}(ku^*[[x]], k^{A^*})$$
which is natural for ku^* algebra maps of k and group homomorphisms of A, which is an isomorphism at the level of varieties.

Proof: First we need a natural map. Choose an orientation of ku. We define
$$\theta : \mathrm{spf}(ku^*BA)(k) \longrightarrow \mathrm{Hopf}_{cts}(ku^*[[x]], k^{A^*})$$
by taking $([\theta(f)](x))(\alpha)$ to be the image of the orientation under the composite
$$ku^*BS^1 \xrightarrow{B\alpha^*} ku^*BA \xrightarrow{f} k.$$

Now we observe that both sides behave well under taking products of abelian groups, so that it suffices to treat the case of cyclic groups. For the codomain we have
$$k^{(C \times D)^*} = k^{C^*} \otimes k^{D^*},$$
and this is a product of Hopf algebras, so the codomain on a product of abelian groups is the product of schemes. For the domain, we have a Künneth map
$$ku^*(B(C \times D)) \longleftarrow ku^*(BC) \hat{\otimes}_{ku^*} ku^*(BD),$$
from the coproduct of topological ku^*-algebras. The main technical ingredient in the present proof is Theorem 1.5.1, which states that the Künneth map induces an isomorphism of varieties. The proof of this will occupy Sections 1.3, 1.4 and 1.5, but for the present we assume the result, so it remains to observe that the map θ factors through the Künneth map of schemes. For this we use the following

diagram, in which we write $\hat{\mathbb{G}}_m(R) = \text{Hopf}_{cts}(ku^*[[x]]$ for brevity.

$$\begin{array}{ccc}
\text{spf}(ku^*BA)(k) & \xrightarrow{\theta} & \hat{\mathbb{G}}_m(k^{A^*}) \\
\downarrow & & \downarrow \\
\text{spf}(ku^*BC \hat{\otimes}_{ku^*} ku^*BD)(k) & & \hat{\mathbb{G}}_m(k^{C^*} \otimes k^{D^*}) \\
=\downarrow & & \downarrow= \\
\{\text{spf}(ku^*BC) \times_{\mathfrak{X}} \text{spf}(ku^*BD)\}(k) & \xrightarrow{\theta \times \theta} & \hat{\mathbb{G}}_m(k^{C^*}) \times_{\mathfrak{X}} \hat{\mathbb{G}}_m(k^{D^*})
\end{array}$$

It thus remains to verify the theorem for cyclic groups. In section 2.2 we use the Gysin sequence to show that if C is cyclic of order n then

$$ku^*(BC) = ku^*[[x]]/([n](x)).$$

Thus
$$\begin{aligned}
\mathfrak{X}(C)(k) &= \text{Rings}_{cts}(ku^*BC, k) \\
&= \{y \in \text{nil}(k) \mid [n](y) = 0\} \\
&= \text{Grp}(C^*, \text{Rings}_{cts}(ku^*[[x]], k)) \\
&= \text{Hopf}_{cts}(ku^*[[x]], \text{Sets}(C^*, k)).
\end{aligned}$$

All these isomorphisms except the last are obvious. The final one is a restriction of the usual one for functions. Indeed if we let

$$f : C^* \longrightarrow \text{Rings}_{cts}(ku^*[[x]], k)$$

and

$$g : ku^*[[x]] \longrightarrow \text{Sets}(C^*, k) = k^{C^*}$$

denote typical functions, the condition that they correspond is that

$$g(p)(\alpha) = f(\alpha)(p).$$

Using the identification $\text{Rings}_{cts}(ku^*[[x]], k)) = \text{nil}(k)$, the condition that f is a group homomorphism is that

$$f(\alpha\beta) = f(\alpha) \odot f(\beta) = f(\alpha) + f(\beta) - vf(\alpha)f(\beta).$$

The condition that g is continuous is that $g(x) \in \text{nil}(\text{Sets}(C^*, k)) = \text{Sets}(C^*, \text{nil}(k))$. The condition that it is a Hopf map is that

$$g(x)(\alpha) + g(x)(\beta) = g(x)(\alpha\beta) + vg(x)(\alpha)g(x)(\beta).$$

$\square$

It is easy to see that the ku^*-algebra

$$ku^*[[e(\alpha) \mid \alpha \in A^*]]/\left(e(\alpha) + e(\beta) = e(\alpha\beta) + ve(\alpha)e(\beta) \mid \alpha, \beta \in A^*\right),$$

represents the functor $\text{Hopf}_{cts}(ku^*[[x]], k^{A^*})$, so the theorem implies most of following.

COROLLARY 1.1.3. *For an abelian group A the ku^*-algebra $ku^*(BA)$ has a subalgebra*

$$ku^*[[e(\alpha) \mid \alpha \in A^*]]/\left(e(\alpha) + e(\beta) = e(\alpha\beta) + ve(\alpha)e(\beta) \mid \alpha, \beta \in A^*\right),$$

and any element of $ku^*(BA)$ has a power which lies in the subalgebra. If A is of rank ≤ 2 the subalgebra is equal to $ku^*(BA)$.

Proof: Suppose $A = C^{(1)} \times C^{(2)} \times \cdots \times C^{(r)}$ with each $C^{(i)}$ cyclic. The theorem shows that $ku^*(BA)$ has the same variety as the tensor product
$$ku^*(BC^{(1)}) \hat{\otimes}_{ku^*} ku^*(BC^{(2)}) \hat{\otimes}_{ku^*} \cdots \hat{\otimes}_{ku^*} ku^*(BC^{(r)})$$
which in turn represents the functor $\mathrm{Hopf}_{cts}(ku^*[[x]], k^{A^*})$. The Künneth theorem together with the fact that $ku^*(BC^{(i)})$ is of flat dimension 1 shows that the tensor product is actually a subalgebra of $ku^*(BA)$. The statement about rank 2 groups follows from the explicit flat resolution of $ku^*(BC^{(1)})$ given by the Gysin sequence (see Section 2.2), together with the fact that $ku^*(BC^{(2)})$ has no $\mathbb{Z}$ or v torsion. $\square$

To see how this Corollary works, note that if C is cyclic, we know $ku^*(BC) = \mathbb{Z}[v][[y]]/([n](y))$, and $y = e(\alpha)$ where α generates C^*. The relation
$$e(\alpha^i) + e(\alpha^j) = e(\alpha^{i+j}) + ve(\alpha^i)e(\alpha^j)$$
implies, by induction, that $e(\alpha^i) = (1 - (1 - ve(\alpha))^i)/v$ from which we conclude that $e(\alpha)$ generates. It also gives $e(\epsilon) + e(\epsilon) = e(\epsilon) + ve(\epsilon)^2$ or $e(\epsilon)(1 - ve(\epsilon)) = 0$; since $1 - ve(\epsilon)$ is a unit in the completed ring we see $e(\epsilon) = 0$. Hence,
$$[n](e(\alpha)) = (1 - (1 - ve(\alpha))^n)/v = e(\alpha^n) = e(\epsilon) = 0,$$
as we know to be correct.

CONVENTION 1.1.4. *In the sequel we generally use the letter y for an Euler class, because of its association with the coordinate of a formal group law.*

1.2. Implications for minimal primes.

Quillen's argument shows that we can deduce the minimal primes of $ku^*(BG)$ from the abelian case provided we know the answer for $ku^*(BA)$ when A is abelian in a sufficiently functorial way. Our proof that $ku^*(BA)$ is a tensor product up to varieties lets us identify its minimal primes.

We begin by considering abelian groups A.

PROPOSITION 1.2.1. *The minimal primes of $ku^*(BA)$ are in bijective correspondence with the set*
$$\{C \mid C \subseteq A \text{ cyclic of prime power order}\} \cup$$
$$\{E(p, A) \mid E(p, A) \text{ is a maximal elementary abelian } p \text{ subgroup}\}$$
of subgroups of A.

Proof: We consider primes according to whether or not they contain v.

Primes not containing v correspond to primes of $K^*(BA)$. Indeed, the primes not containing v are the primes of the localization $ku^*(BA)[v^{-1}]$. Now by 1.1.1,
$$ku^*(BA)[v^{-1}] = K^*(BA) = R(A)^{\wedge}_J[v, v^{-1}],$$
and we know by Segal's work [53] that the minimal primes $\wp_C$ of $R(A)$ correspond to cyclic subgroups C. More precisely, $R(C) = \mathbb{Z}[\eta]/(\eta^n - 1)$, and $ku^*(BC) = ku^*[[y]]/[n](y)$, and we have
$$\wp_C = (\mathrm{res}^A_C)^*(\phi_C(\eta))$$
where ϕ_C is the cyclotomic polynomial, and $vy = 1 - \eta$. Of these primes, only the ones meeting the component of the trivial group survive the process of completion. These are exactly the primes corresponding to subgroups of prime power order.

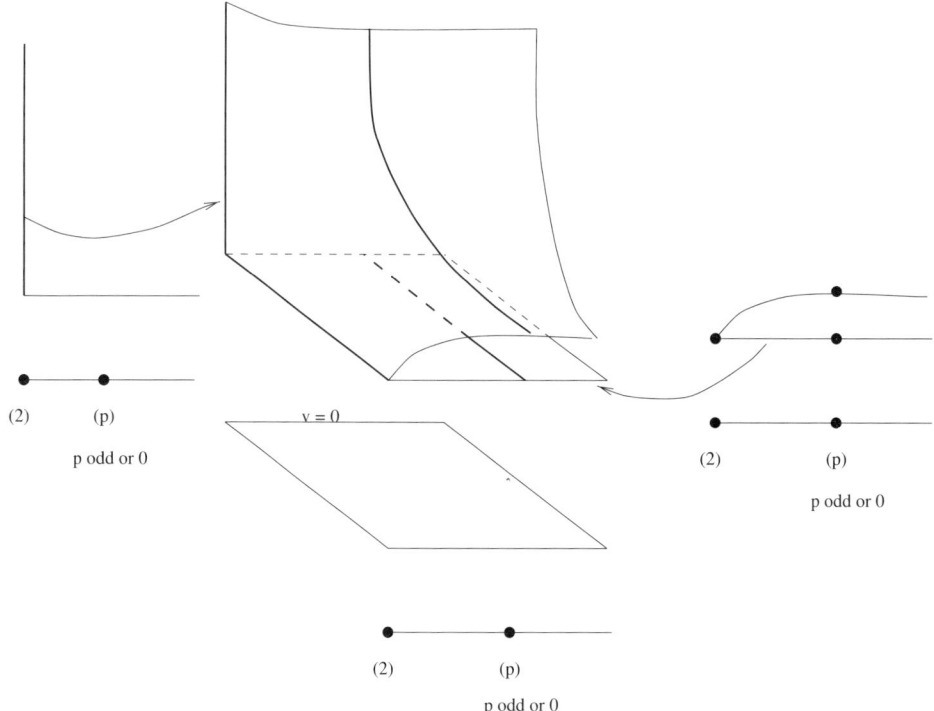

FIGURE 1.1. The varieties of $H\mathbb{Z}^*(BC_2)$, $ku^*(BC_2)$ and $R(C_2)$, and the natural inclusions.

Primes which do contain v correspond to primes of $ku^*(BA)/(v)$, or equivalently, to the corresponding quotient of the subalgebra in 1.1.3, namely

$$ku^*[[e(\alpha) \mid \alpha \in A^*]]/\left(e(\alpha) + e(\beta) = e(\alpha\beta) + ve(\alpha)e(\beta) \mid \alpha, \beta \in A^*\right)/(v) = \mathbb{Z}[[A]].$$

To see this equality, note that the relations become $e(\alpha) + e(\beta) = e(\alpha\beta)$, so that the result is the completed $\mathbb{Z}$ symmetric algebra $\mathbb{Z}[[A]]$ on A^*. The minimal primes of $\mathbb{Z}[[A]]$ are the primes (p), defining affine space over $\mathbb{F}_p$ of dimension $\mathrm{rank}_p(A)$ and the ideal $(e(\alpha) \mid \alpha \in A^*)$, defining $\mathrm{spec}(\mathbb{Z})$.

This identifies all the primes of $ku^*(BA)$, and those minimal within each of the two classes. It remains to identify possible containments between the classes.

None of the second class of primes lie in any of the first class, since they contain v. In the second class $(v, e(\alpha) \mid \alpha \in A^*)$, contains $\wp_1 = (e(\alpha) \mid \alpha \in A^*)$, so is not minimal. Now suppose p divides the group order and consider (v, p). If $A_{(p)}$ is cyclic then $\wp_{C_p}$ lies in (v, p) since it is generated by

$$\phi_p(\eta) = [p](y)/y = \binom{p}{1} - \binom{p}{2}vy + \cdots + \binom{p}{p}(vy)^{p-1}.$$

Similarly, $\wp_{C_{p^k}}$ lies in (v, p) since it is generated by $\phi_{p^k}(\eta) = [p^k](y)/[p^{k-1}](y)$, which is easily seen to be in the ideal (p, v). On the other hand, if $A_{(p)}$ is not cyclic, then (v, p) defines a scheme of dimension $\mathrm{rank}_p(A) > 1$, so it definitely does not contain a prime defining a scheme of dimension 1.

Finally, if p does not divide the group order then (v,p) is certainly not minimal since $ku^*(BA)/(p) = \mathbb{Z}/p[v]$, so that (p) is itself prime. For an algebraic proof note that $[n](y)/y$ has constant term n, and thus if p does not divide n, all irreducible factors are invertible power series mod p, so that the relation $[n](y) = 0$ is equivalent mod p to $y = 0$. □

This analysis also gives an analysis of the primes minimal over each integer prime p.

PROPOSITION 1.2.2. *(i) If p does not divide the group order, (p) is itself prime.*
(ii) If p does divide the group order, there are two primes minimal over (p).

Proof: We have already discussed the case that p does not divide the group order.

Segal shows that the primes $\wp_C$ and $\wp_D$ become equal mod p if the p' parts of C and D are equal. The primes of $R(A)/(p)$ thus correspond to cyclic subgroups of p' order. Completing at J picks out the unique prime $\wp_1$.

It remains only to observe that if p divides the group order the prime (v,p) is always minimal. This is clear if the Sylow p-subgroup is not cyclic, just as before. If it is cyclic of order n, $y \in \wp_1$ and we need only check that $y \notin (v,p)$. However $[n](y) = 0 \mod (p,v)$, so that the defining relation is trivial. □

Here is the answer for a general group:

COROLLARY 1.2.3. *A: Components of $ku^*(BG)$ for a finite group G correspond to*

$$\{C \mid C \text{ cyclic of prime power order }\}/G$$
$$\cup \{A \mid A \text{ maximal amongst elementary abelian subgroups }\}/G,$$

where G acts on these sets by conjugation.

B: Mod p there is one minimal prime if p does not divide the group order. If p does divide the group order there is one minimal prime for each conjugacy class of maximal elementary abelian p-subgroups, and one extra. □

REMARK 1.2.4. (i) Of course $ku^*(BG)/(v)$ is closely related to the integral cohomology, $H\mathbb{Z}^*(BG) = (ku/v)^*(BG)$, and it was shown in [**33**, Appendix E] that their varieties agree. Accordingly the essence of the results is that the variety of ku^*BG contains an amalgamation of the varieties of K^*BG and $H\mathbb{Z}^*BG$:

$$ku = K + H\mathbb{Z}.$$

The interaction takes place at the primes for which the group has rank 1.

(ii) The discussion suggests things about the character of equivariant connective K-theory. We mean an equivariant theory ku_G with the properties (a) if v is inverted, ku_G becomes Atiyah-Segal periodic equivariant K-theory K_G, (b) if we reduce mod v, ku_G has the character of ordinary Borel cohomology and (c) ku_G is complex orientable. Because of complex orientability, the theory formed from K_G by brutally truncating its homotopy does not qualify, but a suitable theory is constructed in [**25**] if G is of prime order and in [**28**] for an arbitrary compact Lie group G.

In $\mathrm{spec}(ku_G^*)$ we expect to find extra components corresponding to cyclic groups with order divisible by more than one prime. In view of Properties (a) and (b), we also expect the corresponding slogan to be

$$ku_G = K_G + H\mathbb{Z}_G,$$

although the term $H\mathbb{Z}_G$ needs elucidation. Since ku_G is not bounded below, the slogan is of less certain value.

Now let us consider a non-abelian example in some detail.

EXAMPLE 1.2.5. The quaternion group Q_8 of order 8.
(i) the components correspond to the 5 conjugacy classes of abelian subgroups and
(ii) mod 2 there are two components.

Proof: The group has three cyclic subgroups $\langle i \rangle, \langle j \rangle$ and $\langle k \rangle$ of order 4, intersecting in the central subgroup of order 2. We have seen that

$$ku^*(BC_2) = \mathbb{Z}[v][[x]]/(x(vx-2))$$

and

$$ku^*(BC_4) = \mathbb{Z}[v][[y]]/(y(vy-2)(v^2y^2 - 2vy + 2))$$

and y restricts to x. Finally, the automorphism group W of order 2 takes η to $\eta^{-1} = \eta^3$, and so $y = (1-\eta)/v$ to $(1-\eta^{-1})/v = (1-(1-vy)^3)/v = v^2y^3 - 3vy^2 + 3y$. To calculate W-invariants we may make use of the fact that for cyclic groups connective K-theory is a subring of periodic K-theory. Now, we begin with $R(C_4)^W$, which is the subring of $R(C_4) = \mathbb{Z}[\alpha]/\alpha^4 = 1$ generated by α^2 and $\alpha + \alpha^{-1}$; this gives the equivariant K-theory by adding the Bott element and $(K^*BC_4)^W$ by completion. We then have

$$ku^*(BC_4)^W = ku^*(BC_4) \cap K^*(BC_4)^W,$$

and it is easy to check this is additively $\mathbb{Z} \oplus \mathbb{Z}_2^{\wedge} \oplus \mathbb{Z}_2^{\wedge}$ in each positive even degree and $\mathbb{Z}_2^{\wedge} \oplus \mathbb{Z}_2^{\wedge}$ in each negative even degree. The inverse limit $\varprojlim_A ku^*(BA)$ can be identified with triples (x_i, x_j, x_k) which restrict to the same element of $ku^*(B\langle -1 \rangle)$ where $x_i \in ku^*(B\langle i \rangle)^W$ and similarly for x_j and x_k. We may thus consider the Quillen restriction map

$$ku^*(BQ_8) \longrightarrow \varprojlim_A ku^*(BA).$$

We calculate $ku^*(BQ_8)$ exactly in Section 2.4, and it may be checked that the restriction map is injective. However the rank of the codomain is greater than that of $ku^*(BQ_8)$ in each even degree.

□

EXAMPLE 1.2.6. The alternating group A_4 on 4 letters.
(i) the components correspond to the 4 conjugacy classes of abelian subgroups and
(ii) mod 2 there are 2 components.
(iii) mod 3 there are 2 components.

1.3. Euler classes and Chern classes.

For any complex stable equivariant cohomology theory we may define Euler classes. We only require the construction for Borel theories defined by $E_G^*(X) = E^*(EG \times_G X)$, but the discussion applies quite generally.

1.3. EULER CLASSES AND CHERN CLASSES.

A complex stable structure is given by natural isomorphisms
$$\widetilde{E}_G^*(S^V \wedge X) \cong \widetilde{E}_G^*(S^{|V|} \wedge X)$$
for simple complex representations V. Here S^V is the one point compactification of V and $|V|$ denotes V with the trivial action. In other words, E_G^* has Thom isomorphisms for topologically trivial bundles. Now for any complex representation V we may consider the inclusion $i_V : S^0 \longrightarrow S^V$, and we may define the Euler class $e(V) \in E_G^{|V|}$ by $e(V) = (i_V)^*(\iota)$, where $\iota \in \widetilde{E}_G^{|V|}(S^V) \cong \widetilde{E}_G^0(S^0)$ corresponds to the unit. It is immediate that $e(V) = 0$ if V contains a trivial summand and that $e(V \oplus W) = e(V)e(W)$.

We shall be particularly concerned with the following four equivariant cohomology theories of a G-space X

1. connective K-theory of the Borel construction $ku^*(EG \times_G X)$
2. integral cohomology of the Borel construction $H^*(EG \times_G X; \mathbb{Z})$
3. periodic K-theory of the Borel construction $K^*(EG \times_G X)$ and
4. periodic equivariant K-theory $K_G^*(X)$

There are natural transformations
$$K_G^*(X)$$
$$\downarrow$$
$$K^*(EG \times_G X) \xleftarrow{\rho_K} ku^*(EG \times_G X) \xrightarrow{\rho_H} H^*(EG \times_G X; \mathbb{Z})$$
between them.

LEMMA 1.3.1. *There are complex stable structures on the above four theories compatible with the displayed natural transformations between them.*

Proof: The three Borel theories are easily seen to be complex stable using the Serre spectral sequence of the fibre sequence $EG \times S^V \longrightarrow EG \times_G S^V \longrightarrow BG$. The complex stable structure on equivariant K-theory comes from equivariant Bott periodicity. The Serre spectral sequence shows that any complex stable structure on ku gives complex stable structures for the other Borel theories. It remains to remark that the image of the Thom class in $\widetilde{K}_G^{|V|}(S^V)$ in $\widetilde{K}^{|V|}(EG \times_G S^V)$ lifts to $\widetilde{ku}^{|V|}(EG \times_G S^V)$. However this is immediate since both Serre spectral sequences collapse. $\square$

CONVENTION 1.3.2. We shall fix a compatible choice of complex stable structures for the rest of the paper. This gives Euler classes $e_{ku}(V)$, $e_H(V)$, $e_K(V)$ and $e_{K_G}(V)$ in the four theories: these classes are central to our calculations and one of the most effective techniques is comparison between the Euler classes of different theories.

The Euler classes in periodic equivariant K-theory are well known, and determined by representation theory.

LEMMA 1.3.3. *For an n-dimensional complex representation V we have*
$$v^n e_{K_G}(V) = \lambda(V),$$
where $\lambda(V) = 1 - V + \lambda^2 V - \cdots + (-1)^n \lambda^n V$. $\square$

It is also useful to know how Euler classes behave under tensor products of representations. In general this is rather complicated, but if G is abelian it is given by an equivariant formal group law in the sense of [**11**]. For equivariant K-theory this is unnecessary because of the direct connection with representation theory, and for Borel theories it is enough to talk about non-equivariant formal group laws. In our case $H\mathbb{Z}$ gives the additive formal group law and ku the multiplicative one.

LEMMA 1.3.4. *If α, β are one dimensional then*
(i) $e_H(\alpha\beta) = e_H(\alpha) + e_H(\beta)$
(ii) *For equivariant K-theory and for connective or periodic K-theory of the Borel construction*

$$e(\alpha\beta) = e(\alpha) + e(\beta) - ve(\alpha)e(\beta)$$

and

$$e(\alpha\beta) = e(\alpha) + \alpha e(\beta). \quad \square$$

The Euler class is the top Chern class, and the other Chern classes also play a role. Suppose then that W is an n-dimensional complex representation. For Borel theories of complex oriented theories we may define Chern classes $c_1^E(W), c_2^E(W), \ldots, c_n^E(W)$ by pullback from the universal classes. Indeed the representation defines a map $W : BG \longrightarrow BU(n)$ and hence

$$E^*[[c_1^E, c_2^E, \ldots, c_n^E]] = E^*(BU(n)) \xrightarrow{W^*} E^*(BG).$$

We define $c_i^E(W) = W^*c_i^E$. For the representation ring we make a similar construction below, and this gives a construction for equivariant K-theory. For the theories that concern us, we choose a complex orientation of ku compatible with the chosen complex stable structure and with the standard orientation on K-theory.

Restriction to the maximal torus identifies $E^*(BU(n))$ as the invariants in $E^*(BT^n)$ under the action of the Weyl group. The universal Chern class $c_j^E \in E^{2j}(BU(n))$ restricts to the j-th symmetric polynomial $\sigma_j(e_1, \ldots, e_n) \in E^{2j}(BT^n)$. To define analogous classes in representation theory we must translate to degree 0, where they should agree with $v^j c_j^K \in K^0(BU(n)) = R(U(n))_J^\wedge$. They are naturally defined in terms of the exterior power operations λ^i.

DEFINITION 1.3.5. *For an n-dimensional complex representation W of G, let*

$$c_j^R(W) = \sum_{i=0}^{j} (-1)^i \binom{n-i}{n-j} \lambda^i(W)$$

LEMMA 1.3.6. *Let W be an n-dimensional representation of G and let E be R or a complex oriented cohomology theory.*
(i) *The natural map $R(G) \longrightarrow K^0(BG)$ sends $c_j^R(W)$ to $v^j c_j^K(W)$.*
(ii) *The Euler class is the top Chern class: $e_E(W) = c_n^E(W)$.*
(iii) *If $c_\bullet^E(W) = 1 + c_1^E(W) + \cdots + c_n^E(W)$ then $c_\bullet^E(V \oplus W) = c_\bullet^E(V)c_\bullet^E(W)$.*
(iv)

$$\lambda^j(W) = \sum_{i=0}^{j} (-1)^i \binom{n-i}{n-j} c_i^R(W)$$

REMARK 1.3.7. (i) A special case of (iii) is $c_\bullet^R(W \oplus \epsilon) = c_\bullet^R(W)$.
(ii) Intuitively, the role of $c_j^R(W)$ is to give the natural expression in W which becomes divisible by exactly v^j in connective k-theory. This observation seems to

be new, though, as noted after the proof of the lemma, they have been known since the 1960's.

Proof: Consider the restriction map

$$\begin{array}{ccccc} R(U(n))^{\wedge}_J & = & Z[\lambda^1 V, \ldots, \lambda^n V, (\lambda^n V)^{-1}]^{\wedge}_J & = & Z[[c_1^R, \ldots, c_n^R]] \\ \downarrow & & & & \\ R(T^n)^{\wedge}_J & = & Z[\alpha_1^{\pm 1}, \ldots, \alpha_n^{\pm 1}]^{\wedge}_J & = & Z[[y_1^R, \ldots, y_n^R]]. \end{array}$$

where V is the defining representation of $U(n)$, α_i is the one-dimensional representation given by projection onto the ith factor, and $y_i^R = 1 - \alpha_i$ is the representation theoretic Euler class. Recall that restriction to the maximal torus induces the map

$$\lambda^i V \mapsto \lambda^i(\alpha_1 + \cdots + \alpha_n) = \sigma_i(\alpha_1, \ldots, \alpha_n).$$

To be compatible with K-theory, we must therefore have

$$\begin{aligned} c_j^R & \mapsto & \sigma_j(y_1^R, \ldots, y_n^R) \\ & = & \sigma_j(1 - \alpha_1, \ldots, 1 - \alpha_n) \\ & = & p_j(\lambda^1, \ldots, \lambda^n) \end{aligned}$$

where $\lambda^i = \lambda^i(\alpha_1 + \cdots + \alpha_n)$ and where p_j is the polynomial which expresses the elementary symmetric functions of the $1 - \alpha_i$ in terms of those of the α_i:

$$\sigma_j(1 - \alpha_1, \ldots, 1 - \alpha_n) = p_j(s_1, \ldots, s_n)$$

where $s_i = \sigma_i(\alpha_1, \ldots, \alpha_n)$. The explicit formula by which we defined c_j^R follows by expanding $\prod_{i=1}^n (x + 1 - \alpha_i)$ in the two obvious ways. Statement (i) follows immediately. Comparison with 1.3.3 proves (ii). Part (iii) follows by well known properties of the symmetric polynomials. Formula (iv) follows from the fact that $t \mapsto 1 - t$ is an involution. □

In these terms, the augmentation ideal $J \subset R(U(n))$ is $J = (c_1^R, \ldots, c_n^R)$ and $\lambda^n \equiv 1 \mod J$, relating the two descriptions of $K^0(BU(n))$ as the power series ring on the c_i^R and as the completion $R(U(n))^{\wedge}_J$.

The Chern classes c_i^R have a long history. See [19] for a comprehensive discussion and bibliography. They are usually defined in terms of the Grothendieck γ-operations. Recall that these are defined by setting

$$\gamma_t(V) = \lambda_{t/(1-t)}(V)$$

and letting $\gamma^i(V)$ be the coefficient of t^i:

$$\gamma_t(V) = \Sigma \gamma^i(V) t^i.$$

In these terms our Chern classes would be $c_i^R(V) = \gamma^i(|V| - V)$, which accords well with our other sign conventions. In particular, $c_{|V|}(V) = e(V)$, the Euler class of V. The other convention which is commonly used, $c_i(V) = \gamma^i(V - |V|)$, differs by $(-1)^i$, so satisfies $c_{|V|}(V) = (-1)^{|V|} e(V)$.

It is perhaps worth finishing by commenting on the ring which models the Euler and Chern classes.

REMARK 1.3.8. A zeroth approximation to the ku-theory Chern subring is given by considering the subring of $K_*^G = R(G)[v, v^{-1}]$ generated by v and the first Chern classes $c_1^{K_G}(V) = v^{-1}(|V| - V)$. This is called the Rees ring $Rees(R(G), J)$, and is a familiar example in algebraic geometry of a blowup algebra. We may consider the ideal I generated by the Chern classes $c_1(V)$ (i.e. the augmentation ideal J in degree -2). The completion of the ring $Rees(R(A), J)$ at I is equal to $ku^*(BA)$ if A is abelian and of rank 1.

For non-abelian groups, it is more natural to consider the subring of $K_*^G = R(G)[v, v^{-1}]$ generated by v and all the Chern classes $c_i^{K_G}(V)$. We call this the modified Rees ring $MRees(G)$. This is equal to the Rees ring if G is abelian. We let I be the ideal generated by the Chern classes, and then the completion of the ring $MRees(G)$ at I is equal to $ku^*(BG)$ if G is of rank 1. However neither $Rees(R(G), J)$ nor $MRees(G)$ has any $\mathbb{Z}$ or v torsion, so neither are good approximations to $ku^*(BG)$ if G is of rank ≥ 2, even if it is abelian.

More generally, $MRees(G)$ is not even a model for the subring of Chern classes. For example we shall see in Chapter 2 that the Euler class $e_{ku}(\tau)$ of the three dimensional simple representation of A_4 is non-zero (since $e_H(\tau) \neq 0$), but maps to zero in periodic K-theory.

Corollary 1.1.3 gives a first approximation if A is abelian: $ku^*(BA)$ has a subring isomorphic to the completion of

$$ku^*[e(\alpha) \mid \alpha \in A^*]/\left(e(\alpha) + e(\beta) = e(\alpha\beta) + ve(\alpha)e(\beta) \mid \alpha, \beta \in A^*\right),$$

and every element of $ku^*(BA)$ has a power in this subring. It is proved in [**27**] that this subring is the universal ring L_A^m for multiplicative A-equivariant formal group laws in the sense of [**11**]. The comparison with the zeroth approximation corresponds to the map from $ku^*(BA)$ to its image in $K^*(BA)$. It is given by the natural map

$$L_A^m \longrightarrow Rees(R(A), J) = MRees(A)$$

which is an isomorphism if A is topologically cyclic and not otherwise. Now $(L_A^m)_I^\wedge = ku^*(BA)$ if A is of rank ≤ 2 and we have equality in even degrees if A is of rank ≤ 3. However the example of elementary abelian groups in Chapter 4 shows that the completion of L_A^m is not a good approximation to $ku^*(BA)$ if A has rank ≥ 4.

Next, we note that $ku^*(BG)$ is often non-zero in odd degrees, even for abelian groups, so none of these models is correct. It is natural to seek a model for the odd dimensional groups by interpreting L_A^m as the ring of functions on $Hom(A^*, \mathbb{G}_m)$ where $\mathbb{G}_m$ is some form of the multiplicative group.

In view of the complexity of the Chern subring of $H^*(BG; \mathbb{Z})$, it seems hopeless to attempt to give a purely algebraic model. A crude approximation is the limit of the Chern rings of abelian subgroups (see Green-Leary [**21**] for more substantial results in this direction).

1.4. Bockstein spectral sequences.

Bockstein spectral sequences have a special property, the 'Tower Lemma', which we shall exploit in our proof that connective K-theory satisfies the Künneth theorem up to varieties. A first step in the proof is the corresponding fact for integral cohomology, which we prove at the end of this section using the Tower Lemma.

Let R be a ring spectrum, and let $a \in R^{-d}$. We then have a cofibre sequence of R-modules

$$\Sigma^d R \xrightarrow{a} R \xrightarrow{\rho} R/a \xrightarrow{\delta} \Sigma^{d+1} R.$$

We organize these into an inverse system

$$\begin{array}{ccccccc} R & \xleftarrow{a} & \Sigma^d R & \xleftarrow{a} & \Sigma^{2d} R & \xleftarrow{a} & \cdots \\ \downarrow & & \downarrow & & \downarrow & & \\ R/a & & \Sigma^d R/a & & \Sigma^{2d} R/a & & \end{array}$$

which we think of as an R/a-Adams resolution of R.

DEFINITION 1.4.1. *The a-Bockstein spectral sequence*, BSS(a), *for R is the spectral sequence of the exact couple obtained by applying $[X, -]^*$ to this diagram.*

PROPOSITION 1.4.2. *BSS(a) is a spectral sequence of R^*-modules with*

$$E_1^{s,t} = (R/a)^{t+sd}(X),$$
$$d_r : E_r^{s,t} \longrightarrow E_r^{s+r,t+1},$$

and $E_\infty^{*,t}$ an associated graded group of $R^t(X)/(a\text{-divisible elements})$.

Proof: The key fact is that there is a map $R \wedge R/a \xrightarrow{\mu_a} R/a$ making the diagram

$$\begin{array}{ccccccc} R \wedge \Sigma^d R & \xrightarrow{1 \wedge a} & R \wedge R & \longrightarrow & R \wedge R/a & \longrightarrow & R \wedge \Sigma^{d+1} R \\ \downarrow & & \downarrow & & \downarrow \mu_a & & \downarrow \\ \Sigma^d R & \xrightarrow{a} & R & \longrightarrow & R/a & \longrightarrow & \Sigma^{d+1} R \end{array}$$

commute. Indeed, the commutativity of the left square implies the existence of μ_a. This means that the maps in the lower cofibre sequence induce R^*-module maps in homotopy. Since the differentials are composites of these maps, the spectral sequence is one of R^*-modules. The description of the E_∞ term is evident from the exact couple. $\square$

Note that we are not asking that R/a be an R-algebra, or that the differentials be derivations. For these properties stronger assumptions on R are required, but we do not need these properties here.

We may succinctly describe the E_1 term as

$$E_1 = (R/a)^*(X) \otimes \mathbb{Z}[a]$$

where $(R/a)^*(X)$ is in filtration 0 (i.e., $E_1^{0,t} = (R/a)^t(X)$) and a is in bidegree $(1, -d)$. Here the R^* action is such that if $x \in R^*X$ is detected by $\bar{x} \otimes a^i$ then ax is detected by $\bar{x} \otimes a^{i+1}$. This follows from the simple fact that if $x : X \longrightarrow R$ lifts to $\bar{x} : X \longrightarrow \Sigma^{id} R$, then $\Sigma^d \bar{x} : \Sigma^d X \longrightarrow \Sigma^{(i+1)d} R$ is a lift of ax.

EXAMPLE 1.4.3. *The Bockstein spectral sequence* BSS(p) *for $H\mathbb{Z}$ is obtained from the cofibre sequence*

$$H\mathbb{Z} \xrightarrow{p} H\mathbb{Z} \longrightarrow H\mathbb{F}_p.$$

It has

$$E_1 = H\mathbb{F}_p^*(X) \otimes \mathbb{Z}[h_0] \Longrightarrow H\mathbb{Z}^*(X)/(p\text{-divisible elements})$$

Here, we violate our abuse of notation since the name h_0 is so well established as the associated graded representative of 2 in an Adams spectral sequence. This is just the ordinary mod p Adams spectral sequence

$$E_2 = \text{Ext}_\mathcal{A}(H\mathbb{F}_p^* H\mathbb{Z}, H\mathbb{F}_p^*(X)) = \text{Ext}_{E[Q_0]}(\mathbb{F}_p, H\mathbb{F}_p^*(X)) \Longrightarrow H\mathbb{Z}^*(X_p^\wedge)$$

which is a much more structured form of the ordinary mod p Bockstein spectral sequence, as its E_∞ term is an associated graded of the object being calculated, unlike the traditional Bockstein spectral sequence which requires that you interpret

E_∞ and the boundaries in each E_r in a special way in order to determine the module being calculated. The multiplicative structure is thereby also made more transparent. Further, it can be compared to other Adams spectral sequences by forgetful functors, e.g., from $E[Q_0, Q_1]$-modules to $E[Q_0]$-modules.

EXAMPLE 1.4.4. The Bockstein spectral sequence BSS(v) for ku is obtained from
$$\Sigma^2 ku \xrightarrow{v} ku \xrightarrow{\rho_H} H\mathbb{Z}$$
Here
$$E_1 = H\mathbb{Z}^*(X) \otimes \mathbb{Z}[v] \Longrightarrow ku^*(X)/(v\text{-divisible elements})$$
This is a mildly disguised form of the Atiyah-Hirzebruch spectral sequence
$$H^*(X, \pi_* ku) \Longrightarrow ku^*(X)$$
since the inverse system is the dual Postnikov tower for ku.

EXAMPLE 1.4.5. The Bockstein spectral sequence BSS(η) for ko is obtained from the cofibre sequence
$$\Sigma ko \xrightarrow{\eta} ko \xrightarrow{c} ku \xrightarrow{r} \Sigma^2 ko$$
in which c is complexification and vr is realification. The reader who has not computed the spectral sequence
$$ku^* \otimes \mathbb{Z}[\eta] = \mathbb{Z}[v, \eta] \Longrightarrow ko^* = \mathbb{Z}[\eta, \alpha, \beta]/(2\eta, \eta^3, \eta\alpha, \alpha^2 - 4\beta)$$
is encouraged to complete this entertaining exercise.

EXAMPLE 1.4.6. The Bockstein spectral sequence obtained from the cofibre sequence
$$S^0 \xrightarrow{2} S^0 \longrightarrow S^0/2 \longrightarrow S^1$$
is a good example of the sufficiency of our minimal hypotheses. Since $(S^0/2)^0 = \mathbb{Z}/2$ and $(S^0/2)^2 = \mathbb{Z}/4$ it is clear that $S^0/2$ is not a ring spectrum. It is also initially surprising that E_∞ is the graded group associated to 2-divisibility, since E_1 is not a $\mathbb{Z}/2$-vector space. However the differentials rid the spectral sequence of these anomalies.

LEMMA 1.4.7. *(The Tower Lemma)* If $x \in R^*(X)$ is detected by $\bar{x} \in E_\infty^{s,t}$ of BSS(a), then there exists $y \in R^*X$ such that $x = a^s y$ and $0 \neq \rho(y) \in (R/a)^*X$.

Proof: By the long exact coefficient sequence, either $\rho(x) \neq 0$ and we are done, since s must then be zero, or $x = ay$ and we may iterate this argument. □

Since the element y of the lemma will be detected in filtration 0, we have the slogan "All a-towers originate on the 0-line in the Bockstein spectral sequence". We emphasize that filtration in BSS(a) exactly reflects divisibility by a.

The tower lemma together with a vanishing line for BSS(p) now allow us to prove that the Künneth homomorphism is a V-isomorphism for abelian groups.

LEMMA 1.4.8. $\widetilde{H\mathbb{Z}}^*(BG)$ is annihilated by $|G|$. □

COROLLARY 1.4.9. If $N = \nu_p(|G|)$, the exponent of p in the order of G, then in BSS(p) for $\widetilde{H\mathbb{Z}}^*(BG)$, $E_N^{**} = E_\infty^{**}$ and $E_\infty^{s,*} = 0$ for $s \geq N$.

Proof: Since $p^N x = 0$ for any p-torsion element $x \in H\mathbb{Z}^*(BG)$, the spectral sequence must collapse at E_N. The Tower Lemma 1.4.7 implies that filtrations $s \geq N$ are 0. $\square$

Now, let $A \cong C^{(1)} \times \cdots \times C^{(r)}$ be a product of cyclic groups.

THEOREM 1.4.10. *The Künneth homomorphism*

$$\kappa_H : H\mathbb{Z}^*(BC^{(1)}) \otimes \cdots \otimes H\mathbb{Z}^*(BC^{(r)}) \longrightarrow H\mathbb{Z}^*(BA)$$

is injective, and every element of $H\mathbb{Z}^*(BA)$ has a power in the image.

Proof: Since $H\mathbb{Z}^* = \mathbb{Z}$ is of homological dimension 1, κ_H is injective.

Now, suppose that $x \in H\mathbb{Z}^*(BA)$. We first show that if $p^i x = 0$, then $x^n \in \mathrm{im}(\kappa_H)$ for some n, using the mod p Bockstein spectral sequence. We may assume for this that A is a p-group. We then have

$$H\mathbb{F}_p^*(BA) = E[x_1, \ldots, x_r] \otimes P[y_1, \ldots, y_r]$$

(unless $p = 2$ and $C^{(i)}$ has order 2, in which case x_i is a polynomial generator and we retain the notation y_i for x_i^2). If $x \in H\mathbb{Z}^*(BA)$ has positive filtration in $\mathrm{BSS}(p)$ then $x^N = 0$ by Corollary 1.4.9, and this is in $\mathrm{im}(\kappa_H)$.

Since $H\mathbb{Z}^* BC = \mathbb{Z}[y]/(ny)$, where $n = |C|$, the image of κ_H is the subalgebra generated by classes which reduce mod p to the y_i's. Thus, if $x \in H\mathbb{Z}^*(BA)$ has filtration 0 in $\mathrm{BSS}(p)$ then $0 \neq \rho_p(x) \in H\mathbb{F}_p^*(BA)$, and $\rho_p(x^p)$ lies in the polynomial subalgebra. Hence there exists $z \in \mathrm{im}(\kappa_H)$ with $x^p - z$ of positive filtration in $\mathrm{BSS}(p)$. Taking m to be a power of p greater than or equal to N, the equation $0 = (x^p - z)^m = x^{pm} - z^m$ then shows that $x^{pm} \in \mathrm{im}(\kappa_H)$.

Finally, for general A, any element $x \in \widetilde{H\mathbb{Z}}^*(BA)$ is the sum of elements of prime power order. Now, if $p_i x_i = 0$ with $(p_i, p_j) = 1$ for $i \neq j$, then $x_i x_j = 0$ for $i \neq j$. Hence, $(x_1 + \cdots + x_k)^N = x_1^N + \cdots + x_k^N$, and the result follows. $\square$

1.5. The Künneth theorem.

We saw in 1.4.10 that if $A = C^{(1)} \times C^{(2)} \times \cdots \times C^{(r)}$ then $H\mathbb{Z}^*(BA)$ is well approximated *as a ring* by the tensor product of the factors. The purpose of this section is to prove the analogous result for $ku^*(BA)$.

For any ring spectrum E there is a Künneth map

$$\kappa_E : E^*(X) \otimes_{E^*} E^*(Y) \longrightarrow E^*(X \times Y).$$

If X and Y are infinite complexes it is appropriate to take into account the skeletal topology. If the target is complete, the map factors through the completed tensor product giving a map

$$\hat{\kappa}_E : E^*(X) \hat{\otimes}_{E^*} E^*(Y) \longrightarrow E^*(X \times Y).$$

This completion has no effect for ordinary cohomology.

THEOREM 1.5.1. *The map*

$$\hat{\kappa}_{ku} : ku^*(BC^{(1)}) \hat{\otimes}_{ku^*} \cdots \hat{\otimes}_{ku^*} ku^*(BC^{(r)}) \longrightarrow ku^*(BA)$$

is injective, and every element of $ku^*(BA)$ has a power in the image. Accordingly $\hat{\kappa}_{ku}$ induces an isomorphism of varieties.

First we remark that this is not a formality. For well behaved theories such as ordinary cohomology and connective K-theory [52] the map κ_E is the edge homomorphism of a Künneth spectral sequence

$$\operatorname{Tor}^{E_*}_{*,*}(E^*(X), E^*(Y)) \Rightarrow E^*(X \times Y).$$

Although the filtration of the spectral sequence is multiplicative, it provides no useful progress towards our result, since the image of κ_E is the bottom filtration rather than the top subquotient.

For integral cohomology the Künneth spectral sequence is a familiar short exact sequence since $H\mathbb{Z}_* = \mathbb{Z}$ is of flat dimension 1. On the other hand $ku_* = \mathbb{Z}[v]$ is of flat dimension 2, so we generally expect a non-trivial differential.

LEMMA 1.5.2. *If C is a cyclic group then $ku^*(BC)$ is of flat dimension 1 as a complete ku^*-module.*

Proof: The ku^*-module $ku^*BS^1 = ku^*[[z]]$ is the completion of a free module and hence flat. The Gysin sequence

$$ku^*(BS^1) \xrightarrow{[n](z)} ku^*(BS^1) \longrightarrow ku^*(BC)$$

gives a flat resolution of length 1. □

COROLLARY 1.5.3. *There is a short exact sequence*

$$0 \to ku^*(BC)\hat{\otimes}_{ku^*} ku^*(X) \longrightarrow ku^*(BC \times X) \longrightarrow \widehat{\operatorname{Tor}}_{ku^*}(ku^*(BC), ku^*\Sigma X) \to 0.$$

□

Next we deal with the easy case of negative codegrees. The good behaviour comes from that of periodic K-theory.

LEMMA 1.5.4. *With A as above, we have the decomposition*

$$R(A) \cong R(C^{(1)}) \otimes \cdots \otimes R(C^{(r)}),$$

and the augmentation ideal $J(A)$ is equal to the sum of the ideals $J(C^{(1)})$, ..., $J(C^{(r)})$.

□

After completion this gives a more directly relevant result.

COROLLARY 1.5.5. *There is an isomorphism*

$$R(A)^{\wedge}_{J(A)} \cong R(C^{(1)})^{\wedge}_{J(C^{(1)})} \hat{\otimes} \cdots \hat{\otimes} R(C^{(r)})^{\wedge}_{J(C^{(r)})},$$

and therefore

$$K^0(BA) \cong K^0(BC^{(1)}) \hat{\otimes} \cdots \hat{\otimes} K^0(BC^{(r)})$$

and κ_K gives an isomorphism

$$K^*(BA) \cong K^*(BC^{(1)}) \hat{\otimes}_{ku^*} \cdots \hat{\otimes}_{ku^*} K^*(BC^{(r)}).$$ □

Using the fact that ku is the connective cover of K we immediately obtain a comparison.

LEMMA 1.5.6. *If $m \geq 0$ and X is a space then*
$$ku^{-2m}(X) = K^{-2m}(X).$$
If X is a connected space (such as BG) then in addition $ku^2(X) \cong \tilde{K}^2(X)$. □

This is enough to show that the Künneth map is an isomorphism in positive degrees.

COROLLARY 1.5.7. *If $m \geq 0$ then*
$$\hat{\kappa}_{ku} : (ku^*(BC^{(1)})\hat{\otimes}_{ku^*} \cdots \hat{\otimes}_{ku^*} ku^*(BC^{(r)}))^{-2m} \xrightarrow{\cong} ku^{-2m}(BA).$$
Furthermore, if $m = 0$ the tensor product may be taken over $ku^0 = \mathbb{Z}$:
$$\hat{\kappa}_{ku} : ku^0(BC^{(1)})\hat{\otimes} \cdots \hat{\otimes} ku^0(BC^{(r)}) \xrightarrow{\cong} ku^0(BA).$$

Proof: We argue by induction on r. Indeed we have a diagram

$$\begin{array}{ccc}
ku^{-2m}(BC^{(1)})\hat{\otimes}ku^0(X) & \xrightarrow{\cong} & K^{-2m}(BC^{(1)})\hat{\otimes}K^0(X) \\
\downarrow & & \downarrow \cong \\
(ku^*(BC^{(1)})\hat{\otimes}_{ku^*}ku^*(X))^{-2m} & \longrightarrow & (K^*(BC^{(1)})\hat{\otimes}_{ku^*}K^*(X))^{-2m} \\
\hat{\kappa}_{ku}\downarrow & & \downarrow \hat{\kappa}_K \\
ku^{-2m}(BC^{(1)} \times X) & \xrightarrow{\cong} & K^{-2m}(BC^{(1)} \times X).
\end{array}$$

It suffices for the first statement to show that $\hat{\kappa}_{ku}$ is an epimorphism since we know by 1.5.3 that it is a monomorphism. If we assume that $\hat{\kappa}_K$ is an isomorphism, this follows from the upper route round the diagram, where the upper right vertical isomorphism follows from Bott periodicity. □

We are now ready to turn to the main proof. First, we have an immediate consequence of the fact that $ku^*(BA)$ is Noetherian.

LEMMA 1.5.8. *There is a bound on the v-power torsion in $ku^*(BA)$. More explicitly, there is a number N_v so that for any $x \in ku^*(BA)$ if $v^{N_v+r}x = 0$ then $v^{N_v}x = 0$. Accordingly the ideal (v^{N_v}) has no v torsion.* □

The main technical ingredient is as follows.

PROPOSITION 1.5.9. *The ideal (v^{N_v}) in $ku^*(BA)$ lies in the image of $\hat{\kappa}_{ku}$:*
$$(v^{N_v}) \subseteq \mathrm{im}(\hat{\kappa}_{ku}).$$

Proof: Note that this is a trivial consequence of 1.5.7 in negative codegrees, so it suffices to consider what happens in codegree $2m$ with $m \geq 0$.

We begin with some results comparing filtrations.

LEMMA 1.5.10. *In $ku^0 BG$ we have the equality*
$$(v)^0 = \hat{J}(G).$$

Proof: Consider the diagram
$$\begin{array}{ccc}
ku^2(BG) & \xrightarrow{v} & ku^0(BG) & \longrightarrow & H^0(BG;\mathbb{Z}) \\
& & \downarrow & & \downarrow \cong \\
& & ku^0(*) & \xrightarrow{\cong} & H^0(*;\mathbb{Z})
\end{array}$$

By definition $\hat{J}(G)$ is the kernel of the left hand vertical, so the result follows from the two isomorphisms and the commutativity of the square. □

There are three filtrations we need to consider on $ku^0(BG)$
(i) the $\hat{J}(G)$-adic filtration, with nth term $\hat{J}(G)^n$
(ii) the skeletal filtration, with nth term
$$Sk^{2n-1} = \ker(ku^0(BG) \longrightarrow ku^0(BG^{2n-1}))$$
(iii) the v-filtration
$$(v)^0 \supseteq (v^2)^0 \supseteq (v^3)^0 \supseteq (v^4)^0 \supseteq \cdots .$$

We summarize some well known relations between the filtrations.

LEMMA 1.5.11. *(i) The $J(G)$-adic and skeletal filtrations on $ku^0 BG$ define the same topology, and the module is complete and Hausdorff for it. Furthermore, the v-topology is finer than the skeletal topology in the sense that $(v^n)^0 \subseteq Sk^{2n-1}$.*
(ii) If $G = A$ is abelian the $J(A)$-adic and skeletal filtrations are equal. In particular, for any n we have $(v^n)^0 \subseteq \hat{J}(A)^n$.

Proof: The Atiyah-Segal completion theorem shows that the $J(G)$-adic and skeletal filtrations on $K^0(BG) = ku^0(BG)$ define the same topology, which is complete and Hausdorff. Now consider

$$\begin{array}{ccc} ku^{2N}(BG) & \xrightarrow{v^N} & ku^0(BG) \\ \downarrow & & \downarrow \\ ku^{2N}(BG^{(2n-1)}) & \xrightarrow{v^N} & ku^0(BG^{(2n-1)}). \end{array}$$

Since $(v^N)^0$ is the image of the top horizontal in the diagram, and since the group $ku^{2N}(BG^{(2n-1)}) = 0$ for $2N > 2n-1$, it follows that $(v^n)^0 \subseteq Sk^{2n-1}$.

If A is abelian, Atiyah [2] shows that the skeletal and $J(A)$-adic filtrations on $K^0(BA) = ku^0(BA)$ agree; this is easily seen for cyclic groups from the Gysin sequence, and it follows in general by the Künneth isomorphism for K^0. □

For cyclic groups we can easily obtain an explicit containment between the topologies in the other direction. Notice that $(v^i)^0$ is not the ith power of an ideal, so the result is not a direct consequence of 1.5.11.

LEMMA 1.5.12. *If C is a cyclic group then $\hat{J}(C)^i \subseteq (v^i)^0$.*

Proof: The ideal $J(C)$ is the principal ideal generated by the degree 0 Euler class vy. □

Now, to continue the proof of 1.5.9, consider the diagram

$$\begin{array}{ccc} (\bigotimes_1^r ku^*(BC^{(i)}))^0 & \xrightarrow{\cong} & ku^0(BA) \\ v^m \uparrow & & \uparrow v^m \\ (\bigotimes_1^r ku^*(BC^{(i)}))^{2m} & \longrightarrow & ku^{2m}(BA) \supseteq (v^{N_v})^{2m} \\ v^{N_v} \uparrow & & \uparrow v^{N_v} \\ (\bigotimes_1^r ku^*(BC^{(i)}))^{2(N_v+m)} & \longrightarrow & ku^{2(N_v+m)}(BA) \end{array}$$

By 1.5.11 $(v^{N_v+m})^0 \subseteq J^{N_v+m}$ for all m. We show that any $s \in (v^{N_v})^{2m}$ lies in the image of the Künneth map. Because of the isomorphism 1.5.7 at the top, $v^m s = \hat{\kappa}_{ku}(\Sigma_\alpha t_{1\alpha} \otimes \cdots \otimes t_{r\alpha})$. Since $v^m s \in (v^{N_v+m})^0 \subseteq J(A)^{N_v+m}$, and since $J(A)^{N_v+m} =$

$\Sigma_{a_1+\cdots+a_r=N_v+m} J_1^{a_1} J_2^{a_2} \cdots J_r^{a_r}$, by 1.5.7 we may suppose $t_{i\alpha} \in J(C^{(i)})^{a_i}$. Since $J(C^{(i)})^{a_i} \subseteq (v^{a_i})^0$ by 1.5.12, we may choose $u_{i\alpha} \in ku^*(BC^{(i)})$ so that

$$t_{1\alpha} \otimes \cdots \otimes t_{r\alpha} = v^{N_v+m} u_{1\alpha} \otimes \cdots \otimes u_{r\alpha}.$$

Taking

$$u = \Sigma_\alpha u_{1\alpha} \otimes \cdots \otimes u_{r\alpha} \in (\bigotimes_1^r ku^*(BC^{(i)}))^{2(N_v+m)}$$

we find

$$s' = \hat{\kappa}_{ku}(v^{N_v} u) = v^{N_v} \hat{\kappa}_{ku}(u) \in (v^{N_v})^{2m}$$

with $v^m(s - s') = 0$. Since v is injective on (v^{N_v}) by choice of N_v we have $s = s'$ as required. This completes the proof of 1.5.9. □

The corresponding result for an ideal generated by a power of p is rather simpler because all comparisons can be done in a single degree.

LEMMA 1.5.13. *For an abelian p-group A, let $\mathcal{E}(A) = (e_{ku}(\alpha) \mid \alpha \in A^*)$. There is a number N_p so that $p^{N_p} \mathcal{E}(A)^{N_v} \subseteq (v^{N_v})$. Accordingly the ideal $p^{N_p} \mathcal{E}(A)^{N_v}$ has no v torsion.*

Proof: If α is of order p^i then $p^i e(\alpha)$ is divisible by v from the relation $0 = e(\alpha^{p^i}) = (1 - (1 - ve(\alpha))^{p^i})/v$. □

Proof of 1.5.1: Now we can complete the proof of the main theorem of the section. Suppose $x \in ku^n(BA)$: we must find a power of x in the image of $\hat{\kappa}_{ku}$. We may suppose $n \geq 2$ by 1.5.7.

If A is not a p-group, the map

$$\bigvee_{p \mid |A|} B(A_{(p)})_+ \xrightarrow{\simeq} BA_+$$

of spaces is a based stable equivalence. It follows that the Künneth map for the decomposition $\Pi_p A_{(p)}$ is an isomorphism. We may choose the decomposition of A as a product of cyclic factors so as to respect this, and it follows that it suffices to deal with the case that A is a p-group.

If $x \in (v)$ then $x^{N_v} \in (v^{N_v})$ and we are done by 1.5.9. Otherwise x is not a multiple of v so that $\rho_H(x) \neq 0$. Hence $\rho(x^q) \in im(\hat{\kappa}_H)$ for some q by 1.4.10, and indeed, we may suppose $\rho_H(x^q) = p(e_H(\alpha_1), e_H(\alpha_2), \ldots, e_H(\alpha_r))$ for some polynomial p with degree $\geq N_v$. Hence $x^q - y$ is divisible by v, where $y = p(e_{ku}(\alpha_1), e_{ku}(\alpha_2), \ldots, e_{ku}(\alpha_r))$, and we have $x^q = y + vx'$. We now show that $(x^q)^{p^M}$ lies in the image of $\hat{\kappa}_{ku}$ for M sufficiently large. Indeed, if $p^M \geq N_v$, we have

$$(x^q)^{p^M} = y^{p^M} + \binom{p^M}{1} y^{p^M-1} vx' + \binom{p^M}{2} y^{p^M-2}(vx')^2 + \cdots$$

$$+ \binom{p^M}{N_v - 1} y^{p^M-N_v+1}(vx')^{N_v-1} + v^{N_v} z.$$

By definition y^{p^M} lies in the image of $\hat{\kappa}_{ku}$, and $v^{N_v} z$ lies in the image of $\hat{\kappa}_{ku}$ by 1.5.9, so it remains to deal with the other terms. By 1.5.13, it suffices to choose M

so that
$$\binom{p^M}{i} \equiv 0 \bmod p^{N_p} \text{ for } 0 < i < N_v.$$

This is easily done by taking $M \geq N_p + \nu_p(N_v!)$, since the binomial coefficient $\binom{p^M}{i}$ is a multiple of p^M divided by $i!$. $\square$

CHAPTER 2

Examples of ku-cohomology of finite groups.

In this chapter we give exact calculations of cohomology rings of some small groups. In Section 2.1 we describe the general procedure for calculation, but roughly speaking it involves using the ordinary cohomology to give the E_2-term of an Adams spectral sequence, an analysis of which gives a generating set. For the multiplicative structure the comparison with periodic K-theory is essential; the Euler and Chern classes play a role in both parts of the process.

The groups we consider are the cyclic groups (Section 2.2), the non-abelian groups of order pq (Section 2.3), the quaternion 2-groups (Section 2.4), the dihedral 2-groups (Section 2.5), and the alternating group A_4 (Section 2.6). All of these are of rank 1 or 2, and it is striking that the answer is so complicated even in such simple cases. The comparison with periodic K theory is much less powerful in higher ranks, and we only treat elementary abelian groups. Even for these, we have to work harder for our results, and accordingly we devote Chapter 4 to this case.

Results for many other groups can be deduced by reducing to the Sylow subgroups or their normalizers. For example, we can compute $ku^*BSL_2(3)$ since we know ku^*BQ_8 and ku^*BC_3. We get $ku^*BSL_2(q)$ ($2q$)-locally and $ku^*B\Sigma_q$ ($q(q-1)$)-locally when q is prime.

From our calculations it will be apparent that ku^*BG is something like a Rees ring with respect to the augmentation ideal. For the cyclic group, it is exactly the Rees ring. However, there are two reasons it is not the Rees ring in general: firstly because non-abelian groups have simple representations of dimension more than 1, and secondly because of torsion.

For example, the representation rings of Q_8 and D_8 are isomorphic as augmented rings, yet their connective K-theories differ, even though they are both generated by Euler classes. One reason is that the exterior powers of their two dimensional simple representations behave differently. This means that their Chern classes also behave differently. To see the relevance of this one may consider the universal case, $U(n)$. In $R(U(n))$, the augmentation ideal is $J = (c_1, \ldots, c_n)$, where the c_i are the Chern classes. The class c_i is divisible by exactly v^i in connective K-theory. This suggests a natural modification of the Rees ring construction which we intend to pursue further elsewhere. The second reason is due to the difference in ranks, leading to v-torsion in $ku^*(BD_8)$. Similarly, the rings ku^*BV, $ku^*BD_{2^n}$, and ku^*BA_4 all contain v-torsion.

Even this is not the whole story. For example, the rings ku^*BV, with V elementary abelian, and ku^*BA_4 contain elements not in the Chern subring. One might hope to explain all this by a topological modified Rees ring construction, whose homotopy is the target of a spectral sequence involving derived functors of the modified Rees ring.

The general pattern that emerges may be summarized by considering the short exact sequence
$$0 \longrightarrow T \longrightarrow ku^*(BG) \longrightarrow Q \longrightarrow 0$$
of $ku^*(BG)$-modules where T is the v-power torsion. Thus Q is the image of $ku^*(BG)$ in $K^*(BG)$, and in many cases (A_4 is the only exception amongst the groups we consider) it is the modified Rees ring of $K^*(BG)$ generated by $1, v$ and the Chern classes of representations with c_i placed in degree $-2i$. Although T is defined as the v-power torsion, when G is a p-group it turns out to be the (p, v)-power torsion. If G is of rank 1, $T = 0$, and if G is of higher rank, its growth rate is polynomial of degree $r - 1$.

2.1. The technique.

Our calculations of $ku^*(BG)$ start from the (ordinary complex) representation ring of G and the mod-p cohomology of BG for each p dividing the order of G. The representation ring $R(G)$ gives the periodic K-theory $K^*(BG) = R(G)^\wedge_J[v, v^{-1}]$ by the Atiyah-Segal Theorem. The mod-p cohomology ring gives the E_2 term of the Adams spectral sequence
$$\mathrm{Ext}_{\mathcal{A}}^{*,*}(H^*(ku), H^*(BG)) \Longrightarrow ku^*(BG)^\wedge_p.$$
Calculation of the E_2 term is generally simple, and the differentials are accessible by various means, primarily the Bockstein spectral sequence $BSS(p)$, stable splittings of the spectra involved, periodicities coming from Thom isomorphisms, and our knowledge of $K^*(BG) = ku^*(BG)[1/v]$. The E_∞ term of the Adams spectral sequence gives an additive generating set for $ku^*(BG)$ and determines much of the ku^*-module structure and some of the multiplicative structure. The relation to periodic K-theory and to cohomology then allows us to complete our calculation of the ring $ku^*(BG)$. The additive generating set given by the Adams spectral sequence gives us control over the order of the torsion and divisibility. The fact that the ku-theory Euler classes and Chern classes specialize to those in cohomology and periodic K-theory is important here.

Since the same methods apply to all the calculations, we discuss the preceding outline of the Adams spectral sequence calculation in more detail before proceeding to the special cases. For the rest of this section all spectra are completed at p. To compute the Adams spectral sequence
$$\mathrm{Ext}_{\mathcal{A}}^{*,*}(H^*(ku), H^*(BG)) \Longrightarrow ku^*(BG)^\wedge_p$$
we need the p-local Adams splitting

(1)
$$ku \simeq l \vee \Sigma^2 l \vee \cdots \vee \Sigma^{2(p-2)} l$$

and the fact that $H^*(l) = \mathcal{A} \otimes_{E(1)} \mathbb{F}_p$, where $E(1) = E[Q_0, Q_1]$, the exterior algebra on the Milnor generators Q_0 and Q_1. A standard change of rings argument then gives
$$\mathrm{Ext}_{\mathcal{A}}^{*,*}(H^*(l), H^*(BG)) = \mathrm{Ext}_{E(1)}^{*,*}(\mathbb{F}_p, H^*(BG)).$$
This is easily calculated since bounded below $E(1)$-modules of finite type are sums of free modules and 'lightning flashes', or 'string modules' [1]. The latter modules are at most one dimensional over $\mathbb{F}_p$ in each degree, and are determined by a finite or infinite 'string'
$$\cdots Q_1 Q_0^{-1} Q_1 Q_0^{-1} Q_1 \cdots$$

2.1. THE TECHNIQUE.

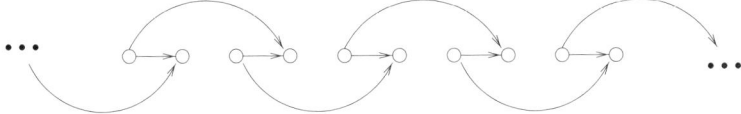

FIGURE 2.1. The bi-infinite string module $\cdots Q_1 Q_0^{-1} Q_1 Q_0^{-1} Q_1 \cdots$ (where Q_0 is denoted by straight, and Q_1 by curved arrows).

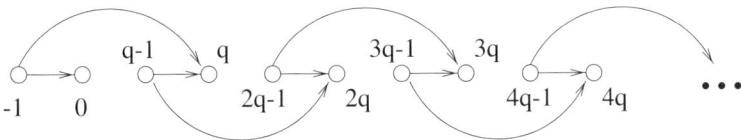

FIGURE 2.2. The module L ($q = 2(p-1)$).

which connects all the elements in the module (see Figure 2.1). Those which are bounded below either start with Q_0^{-1} or Q_1, and those which are bounded above either end with Q_0^{-1} or Q_1, giving four families of finite strings distinguished by length, two bounded below semi-infinite strings, two bounded above semi-infinite strings, and a single string module infinite in both directions. We will let L denote the bounded below module corresponding to the string $Q_0^{-1} Q_1 Q_0^{-1} Q_1 \cdots$, with initial class in degree 0 (see Figure 2.2). This is $\Sigma^{-2} \widetilde{H}^*(BC_2)$ at the prime 2, and is a desuspension of $\widetilde{H}^*(B)$ at odd primes, where B is an indecomposable summand of BC_p.

The ground field $\mathbb{F}_p$ is the string module corresponding to the empty string. This gives the Adams spectral sequence converging to the coefficients,

$$\mathrm{Ext}_{E(1)}^{*,*}(\mathbb{F}_p, \mathbb{F}_p) = \mathbb{F}_p[a_0, u] \Longrightarrow l^* = \mathbb{Z}_p^{\wedge}[u]$$

(see Figure 2.3) with $a_0 \in \mathrm{Ext}^{1,1}$ detecting the map of degree p, (we write h_0 rather than a_0, and v rather than u when $p = 2$), and $u \in \mathrm{Ext}^{1,2p-1}$ detecting v^{p-1}. The Adams spectral sequence must collapse because it is concentrated in degrees divisible by q. When $p = 2$, this gives

$$\mathrm{Ext}_{\mathcal{A}}^{*,*}(H^*(ku), \mathbb{F}_2) = \mathbb{F}_2[h_0, v] \Longrightarrow ku^* = \mathbb{Z}_2^{\wedge}[v]$$

(see Figure 2.3 with $u = v$ and $q = 2$). If $p > 2$, then in terms of the splitting (1), the map $v : \Sigma^2 ku \longrightarrow ku$ is

$$\begin{pmatrix} 0 & 0 & \cdots & 0 & u \\ 1 & 0 & \cdots & 0 & 0 \\ 0 & 1 & \cdots & 0 & 0 \\ \cdot & \cdot & \cdots & \cdot & \cdot \\ \cdot & \cdot & \cdots & \cdot & \cdot \\ \cdot & \cdot & \cdots & \cdot & \cdot \\ 0 & 0 & \cdots & 1 & 0 \end{pmatrix}.$$

It follows that multiplication by v in the Adams spectral sequence

$$\mathrm{Ext}_{\mathcal{A}}^{s,t}(H^*(ku), H^*(\bullet)) \Longrightarrow ku^{-(t-s)}(\bullet)$$

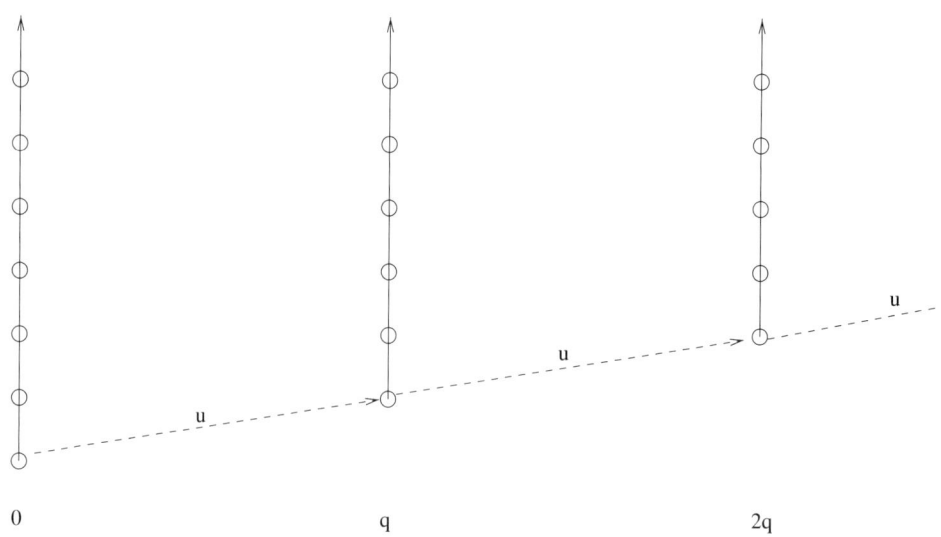

FIGURE 2.3. The Adams spectral sequence for l^*.

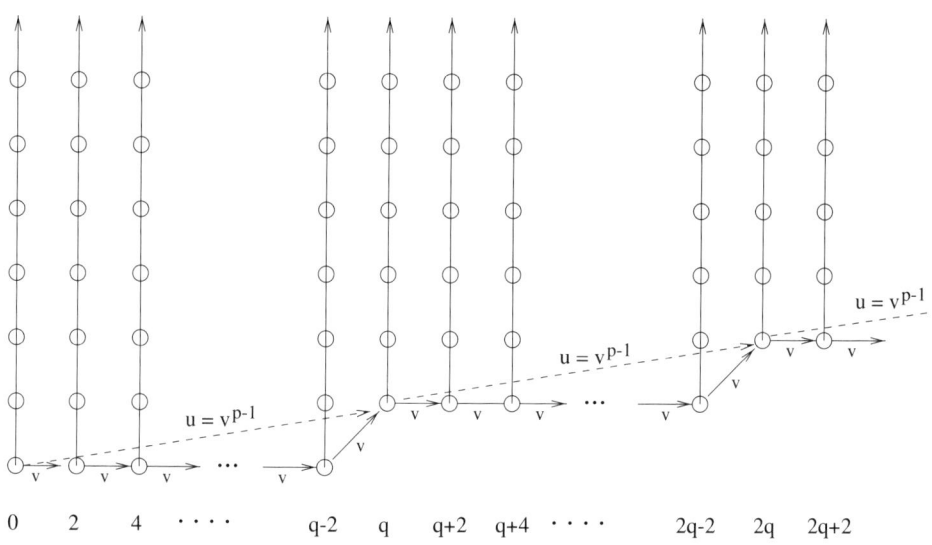

FIGURE 2.4. The Adams spectral sequence for ku^*.

maps each of the first $p-2$ summands isomorphically to the next, preserving filtration, and maps the last summand to the first by multiplication by $v^{p-1} = u$, raising filtrations by 1. For the coefficients ku^*, the result is

$$\mathrm{Ext}_{\mathcal{A}}^{*,*}(H^*(ku), \mathbb{F}_p) = \mathbb{F}_p[a_0, v, u]/(v^{p-1}) \Longrightarrow ku^* = \mathbb{Z}_p^{\wedge}[v]$$

(see Figure 2.4).

The cohomology $\mathrm{Ext}_{E(1)}^{*,*}(\mathbb{F}_p, M)$ of each of the string modules M is readily calculated as a module over $\mathrm{Ext}_{E(1)}^{*,*}(\mathbb{F}_p, \mathbb{F}_p)$. Thus, decomposing $H^*(BG)$ as a sum of indecomposable $E(1)$-modules gives us the E_2 term of the Adams spectral sequence for $ku^*(BG)$. To determine the differentials, the first observation is that stable splittings of BG restrict the possibilities. Stable splittings of $ku \wedge BG$ restrict them even further in some cases (e.g., $G = A_4$ or elementary abelian). The next observation is that the higher Bocksteins in ordinary cohomology imply certain differentials in a manner described below. Having applied these differentials, we can then show, in the cases we study, that the relation to periodic K-theory implies no further differentials can occur. This is in contrast to the case of real connective K-theory ko^*, where a second set of differentials intervenes to impose the relations involving η [6].

There are no convergence problems, despite the fact that $ku^*(BG)$ is unbounded both above and below, since $\widetilde{ku}^i(BG^{(n)})$ is finite if $n > i$, where $BG^{(n)}$ is the n-skeleton of BG. Thus, the inverse system $\{ku^*(BG^{(n)})\}$ is Mittag-Leffler, and we may compute
$$ku^*(BG) = \varprojlim_n ku^*(BG^{(n)}).$$
Convergence also makes it easy to use duality between the ku-cohomology and ku-homology of BG. In particular, information about differentials in the Adams spectral sequence converging to $ku_*(BG)$ can be dualized to yield information about the differentials in the Adams spectral sequence converging to $ku^*(BG)$. This is how splittings of $ku \wedge BG$ are relevant to $ku^*(BG) = \pi_*(F(BG, ku))$.

A situation which will occur repeatedly is the following. Write the cohomology of X as the sum of a module M with no $E(1)$-free summands and a free module over $E(1)$,
$$H^*(X) = M \oplus \bigoplus_\alpha \Sigma^{n_\alpha} E(1).$$
Then Margolis' theorem about Eilenberg-MacLane wedge summands [43] implies that there is a corresponding decomposition,
$$ku \wedge X \simeq \bar{X} \vee \bigvee_\alpha \Sigma^{n_\alpha} H\mathbb{F}_p.$$
Each $E(1)$ in the decomposition of $H^*(X)$ gives rise to $p - 1$ copies of $\mathbb{F}_p$ in filtration 0 of the Adams spectral sequences converging to $ku^*(X)$ and $ku_*(X)$. The decomposition of $ku \wedge X$ shows that these classes do not support any differentials in the spectral sequence for $ku_*(X)$. By S-duality, the classes do not support any differentials in the spectral sequences for $ku_*(DX)$ or $ku^*(X)$ either. Similarly, there are no hidden extensions involving these classes. They detect classes in $ku_*(X)$ and $ku^*(X)$ which are annihilated by p and v.

This duality is involved in the relation between the Bockstein and Adams differentials as well. May and Milgram [46] have shown that above a vanishing line determined by the connectivity of the spectrum, there is a one-to-one correspondence between the h_0-towers in the E_r-term of the Adams spectral sequence $\text{Ext}_A^{*,*}(H^*(X), \mathbb{F}_p) \Longrightarrow \pi_*(X)$ and the r-th term of the Bockstein spectral sequence $BSS(p)$, and that under this correspondence the Adams and Bockstein differentials agree. An h_0-tower is a set $\{h_0^i x | i \geq 0\}$ in which all $h_0^i x \neq 0$, under the obvious equivalence relation of cofinality. We are applying this to the spectral sequence for $l_*(X) = \pi_*(l \wedge X)$ and using the change of rings isomorphism
$$\text{Ext}_A^{*,*}(\mathcal{A} \otimes_{E(1)} H^*(X), \mathbb{F}_p) \cong \text{Ext}_{E(1)}^{*,*}(H^*(X), \mathbb{F}_p)$$

to compute E_2. This introduces a twist to the Bockstein lemma of May and Milgram. The Künneth theorem gives

$$H^*(l \wedge X) \cong H^*(l) \otimes H^*(X) \cong (\mathcal{A} \otimes_{E(1)} \mathbb{F}_p) \otimes H^*(X).$$

In order to convert this to the form needed by the change of rings theorem, we must use the isomorphism

$$\theta : (\mathcal{A} \otimes_{E(1)} \mathbb{F}_p) \otimes H^*(X) \xrightarrow{\cong} \mathcal{A} \otimes_{E(1)} H^*(X)$$

given by $\theta(a \otimes x) = \Sigma a' \otimes a''x$, with inverse $\theta^{-1}(a \otimes x) = \Sigma a' \otimes \chi(a'')x$, where the coproduct on a is $\Sigma a' \otimes a''$. Therefore, when we compute E_2 as $\mathrm{Ext}_{E(1)}$, the Bockstein which determines the Adams differentials is the usual Bockstein conjugated by θ, namely $\widehat{\beta_r} = \theta^{-1}\beta_r\theta$.

The Bockstein spectral sequence for ku is well known [1] to collapse at E_2, with $E_2 = E_\infty$ spanned by the equivalence classes of $Sq^{2n}\iota$, where $\iota \in H^0(ku)$ is the unit. Clearly $Sq^{2n}\iota$ must therefore detect v^n in the Adams spectral sequence for $\pi_*(ku)$, and by the fact that the Adams spectral sequence is multiplicative, we see that a tower corresponding to a class $Sq^{2n}\iota \otimes x$ is v^n times the tower corresponding to x.

Once we have E_∞ of the Adams spectral sequence, we know classes which generate ku^*BG, and the next task is to determine the multiplicative structure. In addition to ad hoc methods, two general techniques are used. First, whenever $v^n : ku^{2n}BG \longrightarrow ku^0BG$ is a monomorphism, we may simply compute in $ku^0BG = K^0BG = R(G)_J^\wedge$. Second, we have a useful consequence of a version of Frobenius reciprocity.

LEMMA 2.1.1. *If V is a representation of G whose restriction to a subgroup H contains a trivial summand, and $\beta \in R(G)$ is induced up from H, then $\beta e_{ku}(V) = 0$.*

Proof: The untwisting isomorphism allows us to calculate

$$\begin{aligned} \beta e &= (\beta' \uparrow_H^G) e \\ &= (\beta' e \downarrow_H^G) \uparrow_H^G \\ &= 0 \end{aligned}$$

where $e = e_{ku}(V)$. $\square$

2.2. Cyclic groups.

The Gysin sequence of the sphere bundle

$$S(\alpha^n) = BC_n \longrightarrow BS^1 \xrightarrow{B(n)} BS^1.$$

associated to the nth tensor power of a faithful simple representation α, splits into short exact sequences because the $[n]$-series is not a zero divisor. This provides presentations of $K^*(BC_n)$ and $ku^*(BC_n)$:

$$K^*(BC_n) = \mathbb{Z}[v, v^{-1}][[e]]/([n](e)) = \mathbb{Z}[v, v^{-1}][[e]]/((1-(1-e)^n))$$

and, sitting inside it,

$$ku^*(BC_n) = \mathbb{Z}[v][[y]]/([n](y)) = \mathbb{Z}[v][[y]]/((1-(1-vy)^n)/v)$$

where $y = e_{ku}(\alpha) \in ku^2(BC_n)$, and $e = ve_K(\alpha) = (1-\alpha) \in K^0(BC_n)$.

2.2. CYCLIC GROUPS.

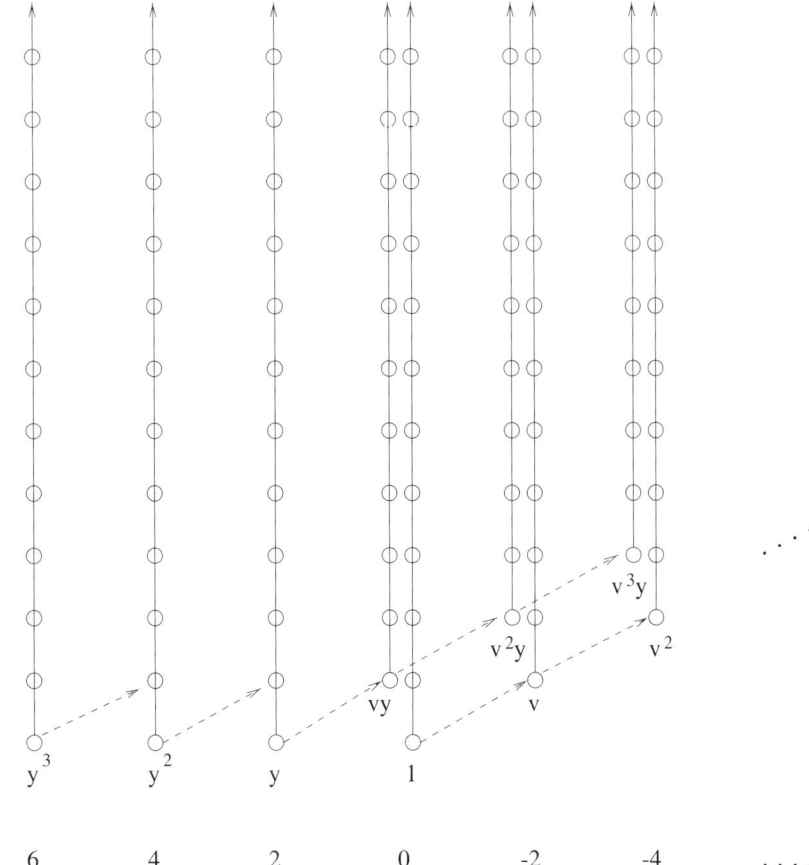

FIGURE 2.5. The Adams spectral sequence for $ku^*(BC_2)$.

This calculation could equally well have been done using Lemma 2.1.1. Take $V = \alpha$ and ρ the regular representation. Then $y = e_{ku}(\alpha)$ satisfies $\alpha = 1 - vy$, and hence

$$0 = y\rho = y \sum_0^{n-1} \alpha^i = \frac{1-(1-vy)}{v} \sum_0^{n-1}(1-vy)^i = \frac{1-(1-vy)^n}{v}.$$

We shall also compute the Adams spectral sequences for the cyclic groups, even though we already have complete information, because we shall need to know how the Adams spectral sequence computation goes for cyclic groups in order to use naturality later.

For the Adams spectral sequence calculation, we may as well restrict attention to the primary case. Let us write $H^*(BC_{p^k}) = E[x] \otimes \mathbb{F}_p[y]$ if $p^k \neq 2$ and $H^*(BC_2) = \mathbb{F}_2[x]$, where $x \in H^1$ and $y \in H^2$. In the latter case, let $y = x^2$. Recall that we have chosen compatible orientations for ku and $H\mathbb{Z}$, so that $y \in ku^2(BC_n)$ maps to $y \in H^2(BC_n)$, making this ambiguity in notation tolerable.

If we start the Adams spectral sequence at E_2 then BC_p and BC_{p^k}, for $k > 1$, appear different. However, if we start the Adams spectral sequence at E_1 using the

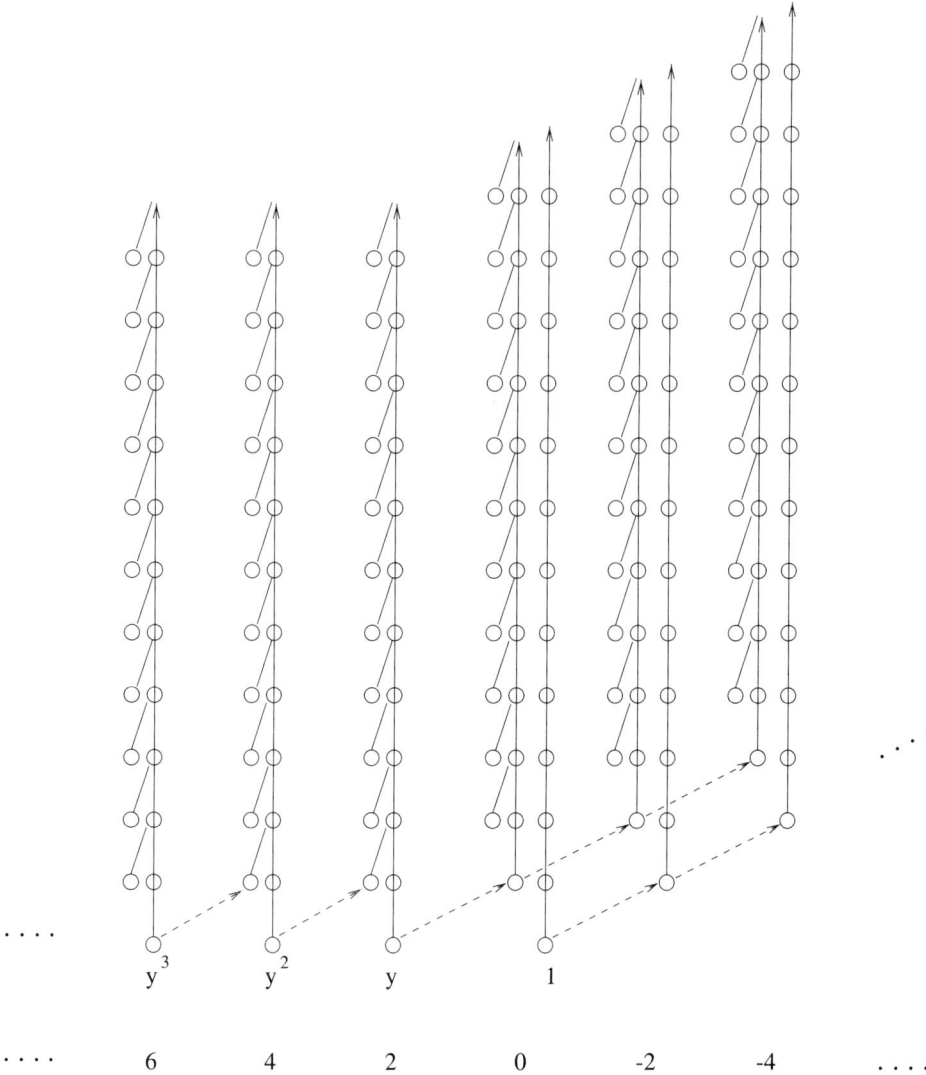

FIGURE 2.6. The Adams spectral sequence for $ku^*(BC_4)$.

Adams resolution obtained by smashing BC_{p^k} with the canonical Adams resolution of a sphere, then we can state the result in a uniform manner.

THEOREM 2.2.1. *In the Adams spectral sequence*

$$\mathrm{Ext}_{\mathcal{A}}^{*,*}(H^*(ku), H^*(BC_{p^k})) \Longrightarrow ku^*(BC_{p^k})_p^\wedge = \mathbb{Z}_p^\wedge[v][[y]]/([p^k]y)$$

$E_1 = E_k$, $E_{k+1} = E_\infty$, $d_k(xy^{n-1}) = h_0^k y^n - h_0^{k-1} v y^{n+1}$ *if* $p = 2$, *and* $d_k(xy^{n-1}) = a_0^k y^n - a_0^{k-1} v^{p-1} y^{n+p-1}$ *if* $p > 2$. *Multiplication by v is a monomorphism on E_∞. The edge homomorphism* $ku^*(BC_{p^k}) \longrightarrow H^*(BC_{p^k})$ *maps* $y \in ku^2(BC_{p^k})$ *to* $y \in H^2(BC_{p^k})$ *and has image* $\mathbb{F}_p[y]$. *As a ku^*-module, $ku^*(BC_{p^k})$ is generated by* $\{y^i \mid i \geq 0\}$.

Proof: It is simplest to describe the calculation completely for the prime 2, followed by the changes needed for odd primes.

With the Adams resolution we have chosen, we have $E_1 = H^*(BC_{2^k}) \otimes \mathbb{Z}[h_0, v]$. We want to show that $d_k(xy^{n-1}) = h_0^k y^n - h_0^{k-1} v y^{n+1}$. The first term follows directly from the cell structure of BC_{2^k}: each even cell of BC_{2^k} is attached to the cell directly below it by a map of degree 2^k. The other term follows by examining the filtrations of the terms in the $[2^k]$-series

$$2^k y - \binom{2^k}{2} v y^2 + \binom{2^k}{3} v^2 y^3 - \cdots - v^{2^k - 1} y^{2^k}.$$

The filtration of

$$\binom{2^k}{i} v^{i-1} y^i$$

is $\nu_2\binom{2^k}{i} + i - 1$, where $\nu_2(n)$ is the exponent of 2 in the prime decomposition of the integer n. It is easy to check that this filtration is k for $i = 1$ or 2, and is greater than k when $i > 2$. The differential $d_k(x) = h_0^k y - h_0^{k-1} v y^2$ follows, and the other d_k's follow from the ring structure. The resulting E_{k+1} is concentrated in even degrees, so the spectral sequence collapses at this point. Further, it is evident that v acts monomorphically. For later reference, let us note that when $k = 1$ this amounts to the calculation

$$\mathrm{Ext}^{*,*}_{E(1)}(\mathbb{F}_2, H^*(BC_2)) = \mathbb{F}_2[h_0, v, y]/(vy^2 - h_0 y)$$

which could also be deduced simply from the operations $Q_0(x) = y = x^2$ and $Q_1(x) = y^2 = x^4$. The 2-primary result is illustrated in Figures 2.5 and 2.6.

When $p > 2$, we have the splitting (1) of ku and a splitting of BC_{p^k}:

(2) $$BC_{p^k} = B_1 \vee \cdots \vee B_{p-1},$$

where B_i has cells in dimensions congruent to $2i - 1$ and $2i$ modulo $2(p-1)$. As an $E(1)$-module, each $H^*(B_i)$ is the odd primary analog of $H^*(BC_{2^k})$. The odd primary calculation of $l^*(B_i)$ is then exactly analogous to the 2-primary calculation of $ku^*(BC_{2^k})$, and the ring structure allows us to assemble these into a calculation of $ku^*(BC_{p^k})$. The details are as follows.

LEMMA 2.2.2. *(Ossa [48]) There is a homotopy equivalence $ku \wedge B_i \simeq ku \wedge \Sigma^2 B_{i-1}$ for $i = 2, \ldots, p-1$. When $k = 1$, $H^*(B_i) = \Sigma^{2i} L$ as an $E(1)$-module, where L is the semi-infinite string module starting in degree -1 as in Section 2.1.*

Proof: The cohomology isomorphism is clear. The Thom complex of the line bundle over $B = BC_{p^k}$ obtained from the representation α is B/B^1, the quotient of B by its 1-skeleton. The ku Thom isomorphism implies that

$$ku \wedge \Sigma^2 B_+ \simeq ku \wedge B^\alpha \simeq ku \wedge B/B^1.$$

Comparing summands in the splitting (2), the result is immediate. $\square$

Now, the calculation of the Adams spectral sequence

$$\mathrm{Ext}^{*,*}_{E(1)}(\mathbb{F}_p, H^*(B_1)) \Longrightarrow l^*(B_1)$$

works exactly as in the 2-primary case. The $[p^k]$-series decomposes exactly as BC_{p^k} does (2), the terms relevant to B_1 being

$$p^k y - \binom{p^k}{p} v^{p-1} y^p + \cdots \pm \binom{p^k}{i(p-1)+1} v^{i(p-1)} y^{1+i(p-1)} \cdots + v^{p^k - 1} y^{p^k}$$

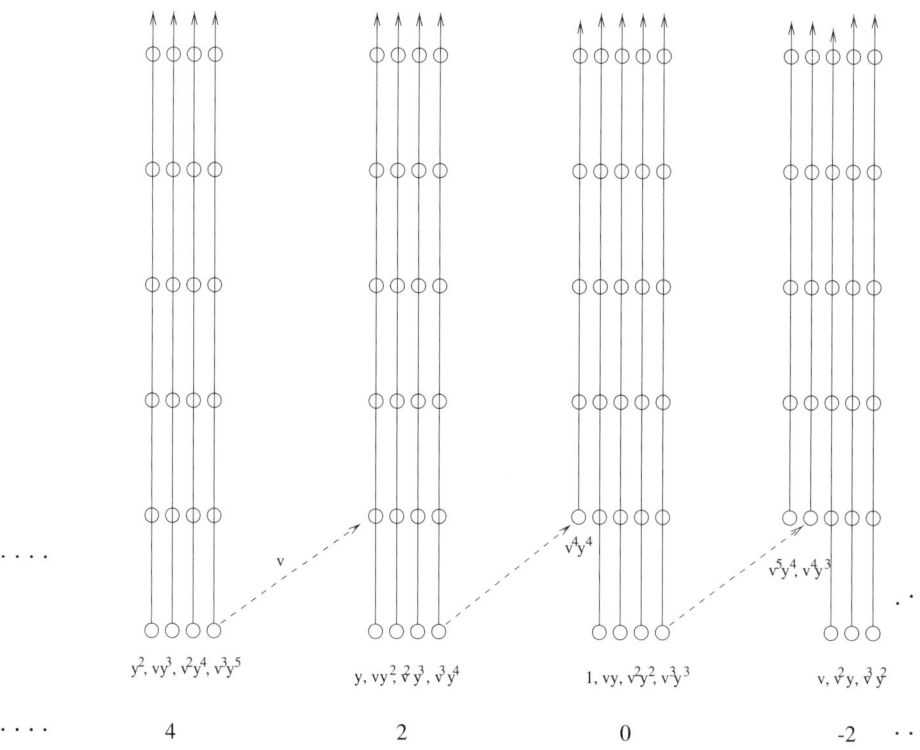

FIGURE 2.7. The Adams spectral sequence for $ku^*(BC_5)$.

and it is a simple matter to check that the lowest filtration terms here are those which correspond to the differential $d_k(x) = a_0^k y - a_0^{k-1} v^{p-1} y^p$. As for $p = 2$, multiplying this relation by powers of y implies the remaining differentials. The rest of the theorem follows exactly as when $p = 2$, though the splitting (1) means that more information is filtered away in the E_2 term. Explicitly,
$$\operatorname{Ext}_{\mathcal{A}}^{s,t}(H^*(ku), H^*(BC_p)) = \mathbb{F}_p[a_0, v, u, y]/(v^{p-1}, uy^p - a_0 y)$$
which is the associated graded group of the quotient of $ku^*[[y]]$ by the p-series
$$v^{p-1} y^p - \binom{p}{p-1} v^{p-2} y^{p-1} + \cdots + py,$$
illustrated in Figure 2.7 when $p = 5$. □

2.3. Nonabelian groups of order pq.

Let q be an odd prime, and let $p \mid q - 1$. We do not assume that p is prime. Let $G_{p,q}$ be the semidirect product of C_q and C_p in which C_p acts upon C_q by an automorphism of order p. For a fixed q, these are exactly the subgroups of the symmetric group Σ_q which sit between $C_q = G_{1,q}$ and its normalizer $N_{\Sigma_q}(C_q) = G_{q-1,q}$. The Atiyah-Hirzebruch-Serre spectral sequence of the extension
$$C_q \xrightarrow{\triangleleft} G_{p,q} \longrightarrow C_p$$

2.3. NONABELIAN GROUPS OF ORDER pq.

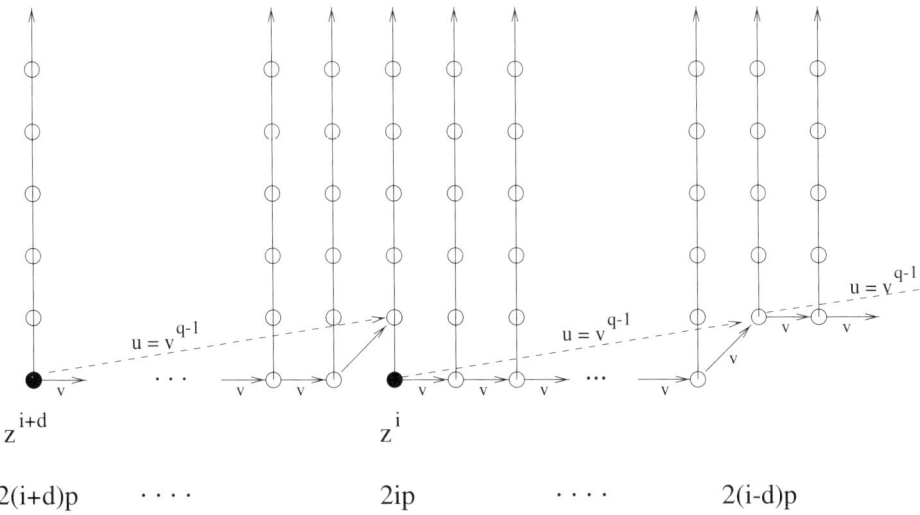

FIGURE 2.8. The $E_2 = E_\infty$ term of the Adams spectral sequence for ku^*B_{ip}.

shows that $BG_{p,q}$ is equivalent to BC_p away from q, and that

$$H^*(BG_{p,q}; \mathbb{F}_q) = H^*(BC_q; \mathbb{F}_q)^{C_p}, \text{ and}$$
$$ku^*(BG_{p,q})_{(q)} = ku^*(BC_q)^{C_p}.$$

The first equality leads to a very nice description of the q-local type of $BG_{p,q}$ in terms of the stable splitting

$$BC_q = B_1 \vee \cdots \vee B_{q-1}.$$

PROPOSITION 2.3.1. *At the prime q,*

$$BG_{p,q} = B_p \vee B_{2p} \vee \cdots \vee B_{q-1}$$

Proof: The action of C_p on C_q is the k-th power map, where k is a primitive p-th root of 1 mod q. The induced action on $H^*(BC_q; \mathbb{F}_q)$ is multiplication by k^i in degrees $2i-1$ and $2i$, so the C_p invariants are exactly the elements in degrees 0 and -1 mod $2p$. The inclusion of the wedge of the B_{ip} into BC_q composed with the natural mapping to $BG_{p,q}$ is therefore an equivalence. $\square$

This gives the additive structure of $ku^*BG_{p,q}$, and the representation ring will give us the multiplicative relations. The group $G_{p,q}$ has p one-dimensional representations $1, \alpha_p, \ldots, \alpha_p^{p-1}$ pulled back from C_p, and has d p-dimensional irreducible representations $\beta_1, \ldots, \beta_d$ induced up from C_q, where $d = (q-1)/p$. Let β be one

of the β_i, and let
$$y = e_{ku}(\alpha_p) \in ku^2(BG_{p,q}), \text{ and}$$
$$z = e_{ku}(\beta) \in ku^{2p}(BG_{p,q}).$$

Let $\rho = 1 + \beta_1 + \ldots + \beta_d$ and observe that if we choose a splitting $C_p \longrightarrow G_{p,q}$ then $\rho \cong 1_{C_p} \uparrow^{G_{p,q}}$, the trivial one-dimensional representation of C_p induced up to $G_{p,q}$.

THEOREM 2.3.2.
$$ku^*(BG_{p,q}) = ku^*[[y,z]]/(yz, [p](y), z\rho).$$

The relation $z\rho = 0$ has the form
$$v^{q-1}z^{d+1} + a_d v^{(d-1)p}z^d + \cdots + a_0 z = 0,$$
where the coefficient $a_0 = \mu q$ for a q-adic unit μ.

Proof: First, observe that under the monomorphism
$$ku^* BG_{p,q} \longrightarrow (ku^* BG_{p,q})_{(p)} \oplus (ku^* BG_{p,q})_{(q)} \longrightarrow ku^* BC_p \oplus ku^* BC_q$$
the element y goes to $(y', 0)$ and z to $(0, z')$ for some y' and z', so that $yz = 0$.

Second, the relation $[p](y) = 0$ follows from $(BG_{p,q})_{(p)} \simeq BC_p$.

Third, we must show that $(ku^* BG_{p,q})_{(q)} = ku^*[[z]]/(z\rho)$. Lemma 2.1.1 implies that $z\rho = 0$, since ρ is induced up from C_p and β contains a trivial representation when restricted to C_p. The splitting 2.3.1 shows that powers of z span $ku^*(BG_{p,q})_{(q)}$ as a ku^*-module because z is detected by a nonzero multiple of y_q^p in $ku^* BC_q$, and powers of y_q^p span the mod q cohomology of $BG_{p,q}$. We have already calculated the Adams spectral sequences for the spectra B_{ip}. (See Figure 2.8.) From these, we see that the first d powers of z are linearly independent, but the $(d+1)$-st must be expressible in terms of the lower ones. Precisely, $\{z, v^p z^2, \ldots, v^{(d-1)p}z^d\}$ span $ku^{2p}(BG_{p,q})_{(q)}$, and therefore $v^{q-1}z^{d+1}$ must be expressible as a linear combination of them with q-adic coefficients. To see that $z\rho$ is such a relation, note that on C_q the β_i restrict to the sums $\alpha_q^{i_1} + \cdots + \alpha_q^{i_p}$, where $\{i_1, \ldots, i_p\}$ is an orbit of C_p acting upon C_q, and α_q is a faithful one-dimensional representation of C_q. Thus, ρ restricts to the regular representation of C_q. Since $\alpha_q^i = 1 - v[i](y_q)$, which has degree i in y_q, we see that ρ has leading coefficient $(v^p z)^d$ as required.

To show that a_0 is as claimed, project onto the summand B_p, and consider the Adams spectral sequence of B_p computed in Section 2.2. (See Figure 2.8). □

When $d = 1, 2,$ or $q - 1$, we can be perfectly explicit. Continue to let α_q be a nontrivial irreducible representation of C_q. We have already computed $ku^* BG_{1,q} = ku^* BC_q$ in the preceding section, but we can give another presentation which is interesting. Let $e_i = e_R(\alpha_q^i) = 1 - \alpha_q^i$. Then $\{e_1, \ldots, e_{q-1}\}$ span the augmentation ideal of $R(C_q)$ and satisfy the relations
$$e_i e_j = e_i + e_j - e_{i+j}.$$
The analogous presentation of $ku^*(BC_q)$ is

(3) $\qquad ku^*(BC_q) = ku^*[[y_1, \ldots, y_{q-1}]]/(vy_i y_j - y_i - y_j + y_{i+j}).$

where $y_j = e_{ku}(\alpha_q^j)$ (compare with 1.1.3).

At the other extreme is $G_{q-1,q} = N_{\Sigma_q}(C_q)$, which is q-locally equivalent to Σ_q.

Theorem 2.3.3.

$$ku^* BN_{\Sigma_q}(C_q) = ku^*[[y,z]]/(yz, [q-1](y), v^{q-1}z^2 - qz)$$

and

$$ku^*(B\Sigma_q)_{(q)} = ku^*[[z]]/(v^{q-1}z^2 = qz)$$

Proof: We need to show that $z\rho = 0$ says $v^{q-1}z^2 = qz$. Now, $z = e_{ku}(\beta)$ and β restricts to the reduced regular representation of C_q. Therefore, $v^{q-1}z$ restricts to $e_1 e_2 \cdots e_{q-1} = e_1 + e_2 + \cdots + e_{q-1} = q - \rho_q$ where ρ_q is the regular representation of C_q. Since $\rho_q^2 = q\rho_q$, we also have $(q - \rho_q)^2 = q(q - \rho_q)$, and the result follows. □

The case $d = 2$ takes a bit more work. When $q \equiv 3 \pmod 4$ the relation $z\rho = 0$ is simple, but when $q \equiv 1 \pmod 4$ it is somewhat trickier. In that case, we express the relation $z\rho = 0$ in terms of the coefficient b in

$$\varepsilon^h = a + b\sqrt{q} \in \mathbb{Q}(\sqrt{q}),$$

where h is the class number of $\mathbb{Q}(\sqrt{q})$ and ε is the fundamental unit (the smallest positive unit greater than 1 in the ring of integers in $\mathbb{Q}(\sqrt{q})$). Unfortunately, no explicit formula for ε in terms of q has been found, though there are simple algorithms ([**39**, Prop 2.4.11]). For example, when $q > 5$, $\varepsilon = (a_1 + b_1\sqrt{q})/2$, where (a_1, b_1) is the positive solution to $a_1^2 - qb_1^2 = -4$ for which b_1 is minimal ([**39**, Rmk 2.4.12 and Lemma 2.4.16]). Note that the coefficients $a, b \in \frac{1}{2}\mathbb{Z}$, so the coefficient $2bq$ in 2 below may well be odd. We exclude $q = 3$ from the following theorem. If $q = 3$ then $p = 1$ and $G_{p,q} = C_3$.

Theorem 2.3.4. When $p = (q-1)/2$, if $y = e_{ku}(\alpha) \in ku^2(BG_{p,q})$ and $z \in ku^{2p}(BG_{p,q})$ are Euler classes as above, and b is the coefficient of $\sqrt{q}$ in ε^h as above, then

(1) when $q \equiv 3 \pmod 4$, $q > 3$,

$$ku^* BG_{p,q} = ku^*[[y,z]]/(yz, [p](y), v^{2p}z^3 + qz)$$

(2) when $q \equiv 1 \pmod 4$,

$$ku^* BG_{p,q} = ku^*[[y,z]]/(yz, [p](y), v^{2p}z^3 - 2bqv^p z^2 + qz)$$

Proof: We need to show that the relation $z\rho$ is as stated. It will suffice to determine the relation satisfied by

$$v^p z = \prod_{i \in (\mathbb{F}_q^\times)^2} (1 - \alpha^i)$$

in $R(C_q)$, where we will now write α for α_q. This is because the representations β_i restrict to orbit sums under the C_p action, and the action of C_p on $C_q \setminus \{0\} = \mathbb{F}_q^\times$ has two orbits, the squares and the non-squares. We may work entirely in $R(C_q) \subset ku^0(BC_q)$ since both $v^{2p} : ku^{2p}(BG_{p,q})_{(q)} \longrightarrow ku^0(BG_{p,q})_{(q)}$ and the natural map $ku^0(BG_{p,q})_{(q)} \longrightarrow ku^0(BC_q)$ are monomorphisms by the splitting in Proposition 2.3.1 and our calculation of the Adams spectral sequence for the cyclic groups in the preceding section.

The notation will be much cleaner if we abuse it by writing z for $v^p z$. The relations we must prove then become

$$z^3 + qz = 0 \quad q \equiv 3 \pmod 4, q > 3, \text{ and,}$$
$$z^3 - 2bqz^2 + qz = 0 \quad q \equiv 1 \pmod 4.$$

Let $N = 1 + \alpha + \alpha^2 + \cdots \alpha^{q-1}$. There are quotient homomorphisms
$$\epsilon : R(C_q) \longrightarrow R(C_q)/(1-\alpha) = \mathbb{Z}, \text{ and}$$
$$\pi : R(C_q) \longrightarrow R(C_q)/(N) = \mathbb{Z}(\zeta) \subset \mathbb{Q}(\zeta), \quad \zeta = e^{2\pi i/q},$$
whose product $R(C_q) \longrightarrow \mathbb{Z} \times \mathbb{Q}(\zeta)$ is a monomorphism.

Let s and n be the sums of the square powers and non-square powers, respectively:
$$s = \sum_{i \in (\mathbb{F}_q^\times)^2} \alpha^i \quad \text{and} \quad n = \sum_{i \in \mathbb{F}_q^\times \setminus (\mathbb{F}_q^\times)^2} \alpha^i.$$

Any element of $R(C_q)$ invariant under C_p can be written as $a + bs + cn$ for some integers a, b, and c. In particular, the image of $R(G_{p,q})$ in $R(C_q)$ is the subring generated by s and n. We shall not use this fact, but it is interesting to note that if we let $q^* = (-1)^{(q-1)/2} q$, then the homomorphism π maps $R(G_{p,q})$ to $\mathbb{Z}(\sqrt{q^*})$, the C_p-invariants in $\mathbb{Z}(\zeta)$:

$$\begin{array}{ccc} R(G_{p,q}) & \longrightarrow & R(C_q) \\ \downarrow & & \downarrow \\ \mathbb{Z}(\sqrt{q^*}) & \subset & \mathbb{Z}(\zeta) \end{array}$$

If k is a non-square modulo q, then $\tau(\alpha) = \alpha^k$ defines an automorphism of $R(C_q)$ which exchanges s and n. (Upon applying π this agrees with the Galois automorphism $\sqrt{q^*} \mapsto -\sqrt{q^*}$.) Let
$$w = \tau(z) = \prod_{i \in \mathbb{F}_q^\times \setminus (\mathbb{F}_q^\times)^2} (1 - \alpha^i).$$

The product $zw = q - N$ since $\epsilon(zw) = 0 = \epsilon(q - N)$ and $\pi(q-N) = q$, while
$$\pi(zw) = \prod_{i=1}^{q-1}(1 - \zeta^i),$$
which equals q by evaluating
$$1 + x + \cdots + x^{q-1} = \prod_{i=1}^{q-1}(x - \zeta^i)$$
at $x = 1$. The proof now divides into two cases depending upon q (mod 4).

First, suppose that $q \equiv 3 \pmod 4$. For any x, $xN = \epsilon(x)N$, so $s(1 + s + n) = p(1 + s + n)$. Hence
$$s^2 + sn = p + (p-1)s + pn.$$
Now $sn = d + e(s+n)$ for some integers d and e, since τ fixes sn. Applying ϵ gives $p^2 = d + 2pe$. Since $q \equiv 3 \pmod 4$, -1 is not a square modulo q, and we may write
$$n = \sum_{i \in (\mathbb{F}_q^\times)^2} \alpha^{-i}.$$

It follows that $d = p$ and hence
$$\begin{aligned} sn &= p + \frac{p-1}{2}(s+n) \\ s^2 &= \frac{p-1}{2}s + \frac{p+1}{2}n \\ n^2 &= \frac{p+1}{2}s + \frac{p-1}{2}n. \end{aligned}$$

Now, if we write $z = a + bs + cn$, then applying ϵ gives $a = -p(b+c)$. Comparing constant terms in

$$\begin{aligned} zw &= (a + bs + cn)(a + cs + bn) \\ &= q - N \\ &= 2p - (s+n) \end{aligned}$$

yields

$$p(b+c)^2 + b^2 + c^2 = 2.$$

Since $q \geq 7$, we have $p > 2$, so $b + c = 0$ and $b^2 + c^2 = 2$. Hence $z = \pm(s-n)$ and it follows easily from the relations satisfied by s and n that $z^3 = -qz$ and that z satisfies no lower degree relation.

Now suppose that $q \equiv 1 \pmod{4}$. From Theorem 2.3.2 we know that $z^3 + a_1 z^2 + a_2 qz$ for some a_1 and a_2 which we will determine by applying π. Let

$$\mathcal{R} = \prod_{i \in (\mathbb{F}_q^\times)^2} (1 - \zeta^i) \quad \text{and} \quad \mathcal{N} = \prod_{i \in \mathbb{F}_q^\times \setminus (\mathbb{F}_q^\times)^2} (1 - \zeta^i).$$

and recall that $\mathcal{RN} = \pi(zw) = q$. The class number formula for $\mathbb{Q}(\sqrt{q})$ implies that if h is the class number and ε is the fundamental unit of $\mathbb{Q}(\sqrt{q})$, then

$$\varepsilon^{2h} = \frac{\mathcal{N}}{\mathcal{R}} = \frac{q}{\mathcal{R}^2}$$

([40, Thm. 218, p. 227]). Thus $\mathcal{R} = \pm\sqrt{q}/\varepsilon^h$. To determine the sign of $\mathcal{R}$, we rewrite it as a product of evidently positive terms. Let r range over $(\mathbb{F}_q^\times)^2$ in the products which follow.

$$\begin{aligned} \mathcal{R} &= \prod_r (1 - \zeta^r) \\ &= \prod_r (-\zeta^{r/2})(\zeta^{r/2} - \zeta^{-r/2}) \\ &= \zeta^{S/2} \prod_r 2i \sin(\pi r/q) \\ &= 2^p i^p \zeta^{S/2} \prod_r \sin(\pi r/q) \\ &= 2^p \prod_r \sin(\pi r/q) \\ &> 0, \end{aligned}$$

where $S = \sum_r r$. (We use the fact that q is prime to get $\zeta^{S/2} = (-1)^{(q-1)/4} = i^p$.) Hence, $\mathcal{R} = \sqrt{q}/\varepsilon^h$. If we write $\varepsilon^h = a + b\sqrt{q}$ with $a, b \in \frac{1}{2}\mathbb{Z}$ then $a^2 - qb^2 = -1$ since the norm of ε is -1 and h is odd ([10, Thm. XI.3, p. 185 and Thm. XI.6, p. 187]). It follows that $\mathcal{R} = bq - a\sqrt{q}$ and this satisfies $\mathcal{R}^2 - 2bq\mathcal{R} + q = 0$. Therefore, $z^3 - 2bqz^2 + qz = 0$ as was to be shown.

Alternatively, if we do not wish to assume that ε has norm -1 and h is odd, we may prove these facts as follows. Let $e = N(\varepsilon)$ be the norm of ε. Then the norm of ε^h, $N(\varepsilon^h) = e^h = (a + b\sqrt{q})(a - b\sqrt{q}) = a^2 - qb^2$. It follows that

$$\mathcal{R} = \frac{\sqrt{q}}{\varepsilon^h} = \frac{\sqrt{q}}{a + b\sqrt{q}} = e^h \sqrt{q}(a - b\sqrt{q}) = e^h(-qb + a\sqrt{q})$$

q	2b	q	2b	q	2b	q	2b
17	2	109	25	229	226	317	5
37	2	113	146	233	3034	337	110671282
41	10	137	298	241	9148450	349	986
61	5	149	5	257	2050	353	7586
73	250	157	17	269	10	373	530
89	106	181	97	277	157	389	130
97	1138	193	253970	281	126890	397	173
101	2	197	2	313	14341370	401	5129602

FIGURE 2.9. Primes $q \equiv 1 \pmod{4}$ for which the coefficient $2b$ in Theorem 2.3.4 is not 1, and the corresponding value of $2b$.

and its conjugate $\mathcal{N} = e^h(-qb - a\sqrt{q})$. Then

$$\begin{aligned}
\mathcal{R}\mathcal{N} &= e^h(-qb + a\sqrt{q})e^h(-qb - a\sqrt{q}) \\
&= q^2b^2 - qa^2 \\
&= -q(a^2 - qb^2) \\
&= -qe^h
\end{aligned}$$

But we know that $\mathcal{R}\mathcal{N} = q$, and it follows that $e^h = -1$. □

REMARK 2.3.5. The proof when $q \equiv 3 \pmod{4}$ is elementary because the coefficients b and c lie on an ellipse with with only two integer points. The same approach for $q \equiv 1 \pmod{4}$ leads to a hyperbola with an infinite number of integer points. This is a reflection of the finitude and infinitude, respectively, of the unit group in the integers of $\mathbb{Q}(\sqrt{q^*})$.

REMARK 2.3.6. As an indication of the likely difficulty of determining the coefficient $-2bq$ as an explicit function of q, Table 2.9 lists those primes $q \equiv 1 \pmod{4}$ up to 401 for which $2b \neq 1$, together with the value of $2b$ for each. These were calculated by using MAGMA ([42]) to calculate the minimal polynomial of $\mathcal{R}$.

REMARK 2.3.7. By analogy with the symmetrical presentation (3) of ku^*BC_q, one might expect to obtain a better presentation of $ku^*BG_{p,q}$ if it were written symmetrically, using the Euler classes of all the β_i. Theorem 2.3.4 shows this is false. The Euler classes $z = e_{ku}(\beta_1)$ and $w = e_{ku}(\beta_2)$ of the two p-dimensional representations each satisfy the relations of Theorem 2.3.4. When $q \equiv 3 \pmod{4}$, $z = -w$, so adjoining w to the presentation does nothing to simplify the relations. When $q \equiv 1 \pmod{4}$,

$$\begin{aligned}
w &= (4b^2q - 1)z - 2bv^p z^2 \text{ and} \\
z &= (4b^2q - 1)w - 2bv^p w^2,
\end{aligned}$$

and we would like $ku^{2p}BG_{p,q}$ to be spanned by z and w. Since z and $v^p z^2$ span, we need $2b$ to be a unit. This may be true in the completed ring, $ku^*BG_{p,q}$, but, if $2b \neq 1$, it will not be true before completion, e.g., in a $G_{p,q}$-equivariant ku^*. If we wish to express everything in terms of z and w, note the relations

$$2bv^p z^2 = (4b^2 q - 1)z - w,$$
$$2bv^p w^2 = (4b^2 q - 1)w - z, \text{ and}$$
$$2bv^p zw = z + w$$

REMARK 2.3.8. It should be clear from the preceding theorem that class field theory is the key to the connective K-theory of the non-abelian groups of order pq. The key relation $z\rho = 0$ can be written as a polynomial in z, a fact which is evident from the Adams spectral sequence, as shown in Theorem 2.3.2. When we do so, we obtain z times a polynomial which defines the C_p fixed subfield of the cyclotomic field $\mathbb{Q}(\zeta_q)$. This is evident from two facts. On the one hand, $\mathcal{R} = \pi(z) = \prod(1-\zeta^i)$, where the product ranges over all i in some C_p-orbit, so that the image of z lies in the C_p-fixed subfield of $\mathbb{Q}(\zeta_q)$. On the other hand, the Adams spectral sequence calculation just cited shows that the minimal polynomial for z has degree $d = (q-1)/p$, so that $\mathcal{R}$ must generate this subfield. (Here we are also using the fact that v^p is a monomorphism on $ku^{2p}BG_{p,q}$, that $v^p z$ lies in $R(G_{p,q})$ and that its powers map to linearly independent elements of $R(C_q)$ and $\mathbb{Q}(\zeta_q)$.)

The extreme cases, $p = 1$ and $p = q-1$ correspond to $\mathbb{Q}(\zeta_q)$ and $\mathbb{Q}$ respectively, but are too simple to show this relation clearly. The case just done, $p = (q-1)/2$, corresponds to the quadratic subfield $\mathbb{Q}(\sqrt{q^*})$, and accordingly, the relation $z\rho = 0$ is z times a quadratic whose coefficients depend in a surprisingly subtle way on q, as shown by the relation $\pi(z) = \sqrt{q}/\varepsilon^h$, for example. The complementary case, $p = 2$, corresponds to the maximal real subfield of $\mathbb{Q}(\zeta_q)$. The corresponding groups $G_{p,q}$ are (non-2-primary) dihedral groups, and we end this section with the explicit description of the first few of these.

By Theorem 2.3.2 we have

$$ku^*(BG_{2,q}) = ku^*[[y,z]]/(yz,[2](y),\rho_{2,q}(z)).$$

with $|z| = 4$. The relation $\rho_{2,q}(z)$ is obtained by inserting powers of v to make the following polynomials homogeneous.

$q = 3$: $z^2 = 3z$
$q = 5$: $z^3 = -5z(1-z)$
$q = 7$: $z^4 = 7z(1-z)^2$
$q = 11$: $z^6 = 11z(1-z)\{(1-z)^3 - z\}$
$q = 13$: $z^7 = -13z(1-z)^2\{(1-z)^3 - 2z\}$
$q = 17$: $z^9 = -17z(1-z)\{(1-z)^6 - 5z(1-z)^3 + z^2\}$
$q = 19$: $z^{10} = 19z(1-z)^2\{(1-z)^6 - 7z(1-z)^3 + 3z^2\}$
$q = 23$: $z^{12} = 23z(1-z)\{(1-z)^9 - 12z(1-z)^6 + 14z^2(1-z)^3 - z^3\}$

There is an obvious guess as to the *form* of these relations for all q, but the coefficients of the terms inside braces are not entirely obvious from this small sample. All of these relations were found simply by computing the minimal polynomial satisfied by $(1-\alpha)(1-\alpha^{-1})$ in $R(C_q)$ using CoCoA [9].

2.4. Quaternion groups.

The simplest p-groups after the cyclic groups are the quaternion groups, as they are the other p-groups with periodic cohomology. As with the cyclic groups, the

Adams spectral sequence shows that $v : ku^*(BQ_{2^n}) \longrightarrow ku^*(BQ_{2^n})$ is a monomorphism, so that $ku^*(BQ_{2^n})$ is a subalgebra of $K^*(BQ_{2^n})$.

REMARK 2.4.1. If G has p-rank 1 for every prime p, it follows that $ku^*(BG)$ has no v-torsion. In all examples we know, if G has p-rank ≥ 2 for some prime p, there is v-torsion in $ku^*(BG)$.

To fix notations, let

$$Q_{2^{n+2}} = <\rho, j \,|\, j^2 = \rho^{2^n}, j^4 = 1, j\rho = \rho^{-1}j> .$$

We will write Q for $Q_{2^{n+2}}$ where it will not be confusing. The mod 2 cohomology rings are

$$H^*(BQ_8; \mathbb{F}_2) = \mathbb{F}_2[x_1, x_2, p_1]/(x_1^2 + x_1x_2 + x_2^2, x_1x_2(x_1+x_2))$$

and, if $n > 1$,

$$H^*(BQ_{2^{n+2}}; \mathbb{F}_2) = \mathbb{F}_2[x_1, x_2, p_1]/(x_1x_2 + x_2^2, x_1^3))$$

with an $(n+2)$-nd Bockstein from H^3 to H^4 [**18**, Prop. VI.5.2, p. 330]. Here $p_1 \in H^4 BQ$ is the first Pontrjagin class of the symplectic representation $Q \longrightarrow S^3 \cong Sp(1)$ obtained by viewing Q as unit quaternions with $\rho = \exp(\pi i/(2^n)) \in \mathbb{C}$.

The representation ring

$$R(Q_{2^{n+2}}) = \frac{\mathbb{Z}[\chi, \psi_0, \psi_1, \ldots, \psi_{2^n}]}{(\chi^2 = 1, \chi\psi_i = \psi_{2^n - i}, \psi_i\psi_j = \psi_{i+j} + \psi_{i-j})}$$

where we let $\psi_{-i} = \psi_i$ and $\psi_{2^n + i} = \psi_{2^n - i}$ in order to describe the product this succinctly. These are defined by

$$\psi_i(\rho) = \begin{pmatrix} \zeta^i & 0 \\ 0 & \zeta^{-i} \end{pmatrix} \qquad \psi_i(j) = \begin{pmatrix} 0 & 1 \\ (-1)^i & 0 \end{pmatrix}$$

$$\chi(\rho) = -1 \qquad \chi(j) = 1$$

where ζ is a 2^{n+1}-st root of unity. Of these, all are irreducible except $\psi_0 = 1 + (\psi_0 - 1)$ and $\psi_{2^n} = \chi + (\psi_{2^n} - \chi)$, which are sums of one-dimensional representations pulled back from the abelianization $Q_{2^{n+2}} \longrightarrow C_2 \times C_2 = <\rho, j \,|\, j^2 = \rho^2 = (\rho j)^2 = 1>$:

$$\psi_0 - 1 \;\longleftarrow\mid\; \widehat{\rho},$$
$$\chi \;\longleftarrow\mid\; \widehat{j}, \text{ and}$$
$$\psi_{2^n} - \chi \;\longleftarrow\mid\; \widehat{\rho j}.$$

Here $\widehat{x}$ denotes the one-dimensional representation with kernel $<x>$. The maximal subgroups are the kernels of these three nontrivial one-dimensional representations,

$$Q_{2^{n+2}}$$
$$\nearrow \quad \uparrow \quad \nwarrow$$
$$Q_{2^{n+1}} \qquad C_{2^{n+1}} \qquad Q_{2^{n+1}}$$
$$\| \qquad\qquad \| \qquad\qquad \|$$
$$<\rho^2, j> \qquad <\rho> \qquad <\rho^2, \rho j>$$

where we let $Q_4 = C_4$ in order to avoid special mention of Q_8.
The following Euler classes will generate ku^*BQ:

$$\begin{aligned} a &= e_{ku}(\psi_0 - 1) &\in& \quad ku^2 BQ \\ b &= e_{ku}(\chi) &\in& \quad ku^2 BQ \\ q_i &= e_{ku}(\psi_i) &\in& \quad ku^4 BQ. \end{aligned}$$

It will also be convenient to define

$$q = q_1 \in ku^4 BQ$$

since the following result shows that we do not need the other q_i to generate ku^*BQ.

LEMMA 2.4.2. For $k = 1, \ldots, 2^n$,

$$q_k = d_k a^2 + \sum_{i=1}^{k} c_{ik} v^{2i-2} q^i$$

where $c_{kk} = (-1)^k$, $c_{1k} = k^2$, and

$$d_k = \begin{cases} -1 & \text{if } k \equiv 2 \pmod{4} \\ 0 & \text{otherwise.} \end{cases}$$

LEMMA 2.4.3. The regular representation of $Q_{2^{n+2}}$,

$$\begin{aligned} \rho_{n+2} &= \psi_0 + 2(\psi_1 + \cdots + \psi_{2^n-1}) + \psi_{2^n} \\ &= v^{2^{n+1}} q^{2^n} + \left(\sum_{i=1}^{2^n-1} f_i v^{2i} q^i \right) + 2^{n+2} \end{aligned}$$

for integers f_i.

REMARK 2.4.4. It is somewhat surprising that the regular representation can be written as a polynomial in the single Euler class $q = e_{ku}\psi_1$.

It is more surprising that the the q_k are closely related to the Chebyshev polynomials. The analogous Euler classes d_k for the the dihedral groups are exactly the Chebyshev polynomials and therefore have many interesting properties, the most remarkable being the composition property $d_k(d_j) = d_{kj}$ (see 2.5.3).

The regular representation ρ_{j+2}, for $j < n$, can be computed by

$$\rho_{j+2} = \frac{q_{2^j+1} - q_{2^j-1}}{q}.$$

In general, the differences $(q_{2^j+i} - q_{2^j-i})/q$ are interesting multiples of ρ_{j+2}.

THEOREM 2.4.5. $ku^*(BQ_{2^{n+2}}) = (ku^*[a,b,q]/I)^\wedge_J$ where $|a| = |b| = 2$ and $|q| = 4$, and where I is the ideal

$$\begin{aligned} I = \quad &(q\rho, va^2 - 2a, vb^2 - 2b, \\ &a^2 - vaq, b^2 - (vbq + q_{2^n - 1} - q), \\ &ab - (b^2 - q_{2^n})) \end{aligned}$$

The natural map $ku^*BQ \longrightarrow H^*BQ$ sends a to x_1^2, b to x_2^2, and q to p_1.

Since Q_8 is both smaller and more symmetric than the other quaternion groups, its connective K-theory has an especially simple form.

THEOREM 2.4.6. $ku^*(BQ_8) = (ku^*[a,b,q]/I)_J^\wedge$ where $|a| = |b| = 2$ and $|q| = 4$, and where I is the ideal
$$\begin{aligned}I = (&v^4q^3 - 6v^2q^2 + 8q,\\ &va^2 - 2a, vb^2 - 2b,\\ &a^2 - vaq, b^2 - vbq,\\ &ab - (vaq + vbq + v^2q^2 - 4q))\end{aligned}$$

The natural map $ku^*BQ_8 \longrightarrow H^*BQ_8$ sends a to x_1^2, b to x_2^2, and q to p_1.

To follow the proofs, it will help to have some relations in front of us. Computing Euler classes in representation theory, we find that

$$\begin{aligned}va &= 2 - \psi_0\\ vb &= 1 - \chi\\ v^2q &= 2 - \psi_1\\ v^2q_i &= \begin{cases} 2 - \psi_i & i \text{ odd}\\ \psi_0 - \psi_i & i \text{ even}\end{cases}\end{aligned}$$

and therefore

$$\begin{aligned}\psi_0 &= 2 - va\\ \chi &= 1 - vb\\ \psi_i &= \begin{cases} 2 - v^2q_i & i \text{ odd}\\ 2 - v^2q_i - va & i \text{ even}\end{cases}\\ \psi_{2^n} &= (2 - va)(1 - vb)\end{aligned}$$

(The final relation follows from $\psi_{2^n} = \chi\psi_0$.)

We now proceed to prove Theorems 2.4.5 and 2.4.6. We first use the Adams spectral sequence to show that there is no v-torsion in ku^*BQ. This will allow us to prove Lemmas 2.4.2 and 2.4.3, and show that the relations in Theorem 2.4.5 hold, by calculating in the representation ring. We can then combine these lemmas with the Adams spectral sequence to produce an additive basis (Lemma 2.4.10), from which we can finish the proof of the theorems by showing that the generators and relations we specify there are complete.

LEMMA 2.4.7. ku^*BQ is concentrated in even degrees and has no v-torsion.

Proof: There is a stable splitting due to Mitchell and Priddy ([47])
$$BQ_{2^{n+2}} \simeq (BSL_2(\ell))_{(2)} \vee X \vee X$$

where $X = \Sigma^{-1}BS^3/BN$ and N is the normalizer of a maximal torus in S^3. Here ℓ is any prime such that the power of 2 dividing $\ell(\ell^2 - 1) = |SL_2(\ell)|$ is exactly 2^{n+2}. The cohomology of X is $\mathbb{F}_2$ in each codegree congruent to 1 or 2 (mod 4), connected by Sq^1, while that of $(BSL_2(\ell))_{(2)}$ is $\mathbb{F}_2$ in each codegree congruent to 3 or 4 (mod 4), connected by the Bockstein β_{n+2}.

Since $\widetilde{ku}^0(BQ_8) = \widetilde{K}^0(BQ_8) = (\mathbb{Z}_2^\wedge)^4$, we must have
$$\widetilde{ku}^0 X = \mathbb{Z}_2^\wedge \quad \text{and} \quad \widetilde{ku}^0(BSL_2(3)_{(2)}) = (\mathbb{Z}_2^\wedge)^2.$$

The cohomology of X is free as an $\mathbb{F}_2[p_1]$-module, and as an $E[Q_0]$-module. We may choose the splitting so that the E_2 term of the Adams spectral sequence for the two summands X are free $\mathbb{F}_2[v,p_1]$-modules on x_1^2 and x_2^2, respectively. By sparseness, no differentials are possible. It follows that there is no v-torsion in ku^*X. Further,

2.4. QUATERNION GROUPS.

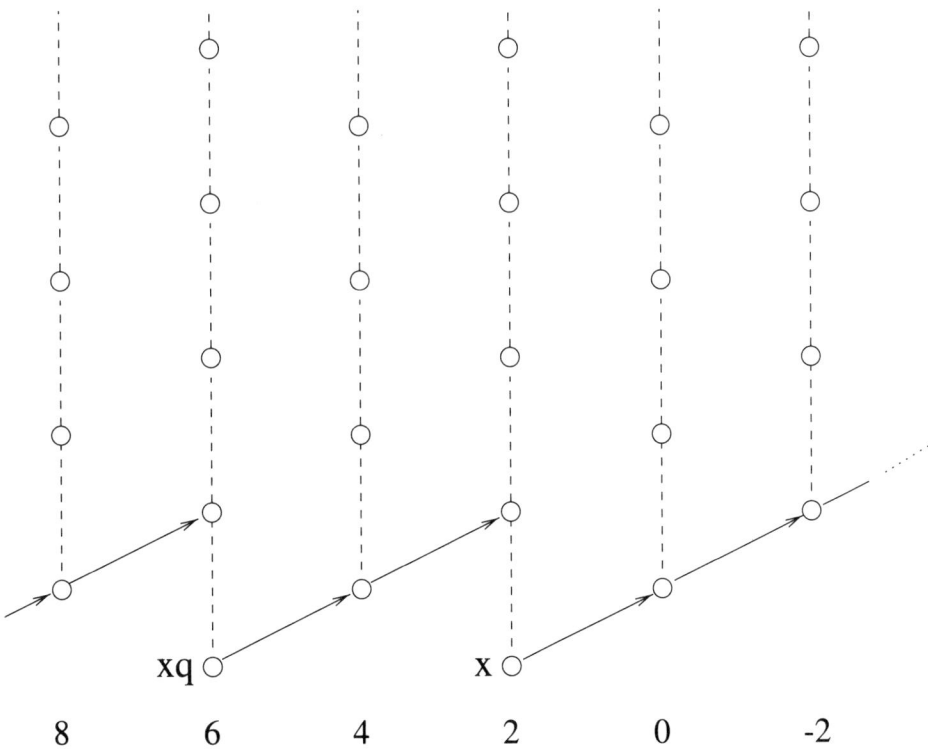

FIGURE 2.10. $E_2 = E_\infty$ in the Adams spectral sequence for ku^*X. (Here x stands for either a or b.)

since $\widetilde{ku}^0 X = \mathbb{Z}_2^\wedge$, twice any element must be detected by v^2 times the element in the same filtration two codegrees higher. See Figure 2.10, in which we continue the practice of showing multiplication by v only for the bottom class of a tower, in order to avoid clutter.

The mod 2 cohomology of the other summand, $BSL_2(\ell)_{(2)}$, is trivial as a module over $E(1)$, so the Adams spectral sequence has

$$\begin{aligned} E_2 &= H^*(BSL_2(\ell)_{(2)}) \otimes \mathbb{F}_2[h_0, v] \\ &= \mathbb{F}_2[h_0, v, p_1] \otimes E[z] \end{aligned}$$

where z generates $H^3 BSL_2(\ell)$. By the correspondence between Adams and Bockstein differentials, the Bockstein $\beta_{n+2}(z) = p_1$ implies an Adams differential

$$d_{n+2}(z) = h_0^{n+2} p_1 + vx$$

for some x. This is not 0, so is not a zero-divisor, so

$$E_{n+3} = \mathbb{F}_2[h_0, v, p_1]/(h_0^{n+2} p_1 + vx)$$

Since this is concentrated in even degrees, $E_{n+3} = E_\infty$. Since v does not divide $h_0^{n+2} p_1$, v is not a zero-divisor on E_∞, and hence also not on $ku^*(BSL_2(\ell))_{(2)}$. □

Now we can show that a relation holds in ku^*BQ by showing that it holds after multiplication by a suitable power of v. We use this to get two relations we need now.

LEMMA 2.4.8. *The relations* $va^2 = 2a$ *and* $a^2 = vaq$ *hold in* ku^*BQ.

Proof: From $\psi_0^2 = 2\psi_0$ we get that $v^2a^2 = 2va$, and from $\psi_0\psi_1 = 2\psi_1$, we find that $v^3aq = 2va = v^2a^2$. $\square$

Proof of Lemma 2.4.2: If $k = 1$ the statement is trivially true. Assume for induction that it holds for k. If k is even, the relation $\psi_1\psi_k = \psi_{k-1} + \psi_{k+1}$ says that
$$(2 - v^2q)(2 - va - v^2q_k) = 2 - v^2q_{k-1} + 2 - v^2q_{k+1}.$$
Cancelling the constant term, replacing $2va$ by v^2a^2, dividing by v^2 and isolating the term q_{k+1}, we find that
$$\begin{aligned} q_{k+1} &= 2q_k + 2q - v^2qq_k - q_{k-1} \\ &= (-v^2)^k q^{k+1} + d_k a^2(2 - v^2q) + P \end{aligned}$$
for some polynomial P of degree k in q. Since $a(2 - v^2q) = 0$, this has the form required. Similarly, if k is odd, using $q_0 = 0$ when $k = 1$, we find
$$\begin{aligned} q_{k+1} &= 2q_k + 2q - v^2qq_k - a^2 - q_{k-1} \\ &= (-v^2)^k q^{k+1} + (-1 - d_{k-1})a^2 + P \end{aligned}$$
for some polynomial P of degree k in q, from which the claimed relation is immediate. These inductive relations easily imply that $c_{1,k} = k^2$. $\square$

Proof of Lemma 2.4.3: This is now a simple calculation using the expression of the ψ_k in terms of a and q_k and Lemma 2.4.2. $\square$

LEMMA 2.4.9. *The relations* $q\rho = 0$, $vb^2 = 2b$, $ab = b^2 - q_{2^n}$ *and* $b^2 = vbq + q_{2^n-1} - q$, *hold in* ku^*BQ.

Proof: Since ρ is induced up from the trivial subgroup, it will be annihilated by any Euler class such as q. Since $\chi^2 = 1$, we see that $vb^2 = 2b$. The Euler classes of the one-dimensional representations are a, b and
$$\begin{aligned} e_{ku}(\psi_{2^n} - \chi) &= e_{ku}(\chi(\psi_0 - 1)) \\ &= e_{ku}(\chi) + \chi e_{ku}(\psi_0 - 1) \\ &= b + (1 - vb)a \\ &= a + b - vab, \end{aligned}$$
so that $\psi_{2^n} = \chi + (\psi_2^n - \chi)$ gives
$$\begin{aligned} q_{2^n} &= e_{ku}(\psi_{2^n}) = e_{ku}(\chi)e_{ku}(\psi_{2^n} - \chi) \\ &= b(a + b - vab) = ab + b^2 - vab^2 = b^2 - ab \end{aligned}$$
and hence $ab = b^2 - q_{2^n}$. Finally, from $\psi_{2^n-1} = \chi\psi_1$, we find $(2 - v^2q_{2^n-1}) = (1 - vb)(2 - v^2q)$, or $b^2 = vbq + q_{2^n-1} - q$. $\square$

LEMMA 2.4.10. *As a module,*

(1) ku^2BQ is the free $\mathbb{Z}_2^\wedge$-module on $\{a, b, vq, \ldots, v^{2^{n+1}-1}q^{2^n}\}$.
(2) ku^4BQ is the free $\mathbb{Z}_2^\wedge$-module on $\{vaq, vbq, q, \ldots, v^{2^{n+1}-2}q^{2^n}\}$.
(3) If $i > 4$ then multiplication by q is an isomorphism from $ku^{i-4}BQ$ to $ku^i BQ$.
(4) If $i < 2$ then multiplication by v is an isomorphism from $\widetilde{ku}^{i+2}BQ$ to $\widetilde{ku}^i BQ$.

Proof: First, observe that multiplication by v is an isomorphism in the Adams spectral sequence in the range indicated, and similarly for multiplication by q. Hence, the same is true in ku^*BQ. Our Adams spectral sequence calculation (Lemma 2.4.7) shows that $ku^*BSL_2(\ell)_{(2)}$ is a quotient of $\mathbb{Z}_2^\wedge[v,q]$ and Lemma 2.4.3 shows that the relation $q\rho = 0$ leaves $\widetilde{ku}^{2i}BSL_2(\ell)_{(2)}$ of rank $2^n - 1$, the same as the rank of $\widetilde{K}^{2i}BSL_2(\ell)_{(2)}$, so there can be no other relation among the powers of q. Now, each X summand contributes a $\mathbb{Z}_2^\wedge$ in each even degree, generated by a and b in degree 2, and by vaq and vbq in degree 4. $\square$

Proof of Theorem 2.4.5: To finish, we need only verify that the relations we have found are a complete set of relations. For this, it is enough to show that the products of v, a, b, and q with elements of the basis found in Lemma 2.4.10 can again be expressed in terms of that basis by means of these relations. Lemma 2.4.10 also shows that we need only consider v times ku^4 and ku^6, and a, b, and q times ku^2 and ku^4. These verifications are then routine once one notices the following consequences of the relations:

$$\begin{aligned}
v^2 bq &= v(b^2 - q_{2^n-1} + q) = 2b - vq_{2^n-1} + vq, \\
v^{2i} aq^{i+1} &= 2v^{2i-2}aq^i, \\
v^{2i} bq^{i+1} &= 2v^{2i-2}bq^i + v^{2i-1}q^{i+1} - v^{2i-1}qq_{2^n-1},
\end{aligned}$$

and $v^{2^{n+1}}q^{2^n+1}$ is a polynomial in v and q of degree 2^n in q. $\square$

Proof of Theorem 2.4.6: We need only show that $q_2 = 4q - v^2q^2 - a^2$, and this follows from $\psi_1^2 = \psi_0 + \psi_2$ as in Lemma 2.4.2. $\square$

We obtain as a corollary, 2-primary information about $SL_2(\ell)$ and complete information about $SL_2(3)$.

COROLLARY 2.4.11. $ku^*(BSL_2(\ell)_{(2)}) = (ku^*[q]/(q\rho))_J^\wedge$ where $|q| = 4$, and ρ is the polynomial in q given in Lemma 2.4.3.

THEOREM 2.4.12. $ku^*(BSL_2(3)) = (ku^*[y, q]/(yq, [3](y), v^4q^3 - 6v^2q^2 + 8q))_J^\wedge$ where $|y| = 2$ and $|q| = 4$.

Proof: The Atiyah-Hirzebruch spectral sequence of $Q_8 \longrightarrow SL_2(3) \longrightarrow C_3$ shows that $ku^*BSL_2(3)_{(3)} = ku^*BC_3$. We have just computed $ku^*BSL_2(3)_{(2)}$. Since y and q each map trivially to opposite localizations, $yq = 0$. $\square$

2. EXAMPLES OF ku-COHOMOLOGY OF FINITE GROUPS.

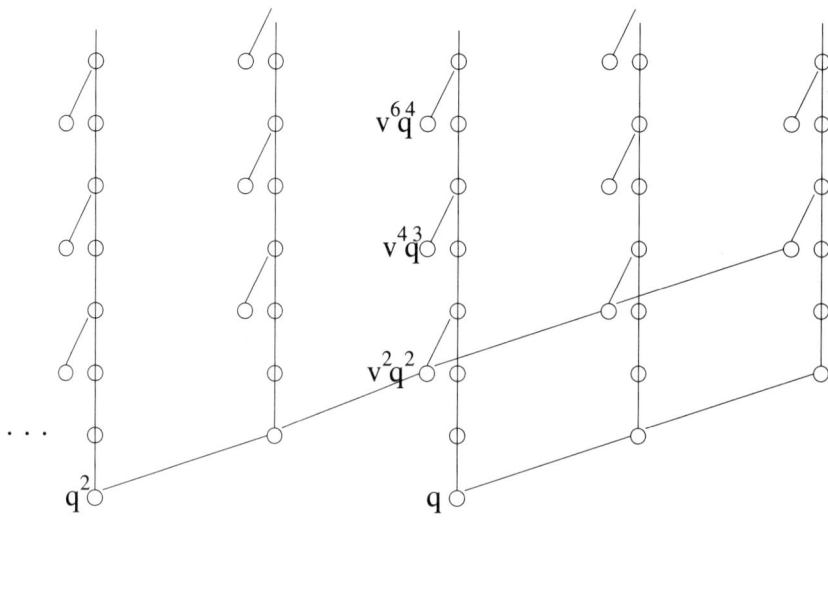

FIGURE 2.11. $E_4 = E_\infty$ in the Adams spectral sequence for $ku^*(BSL_2(3)_{(2)})$.

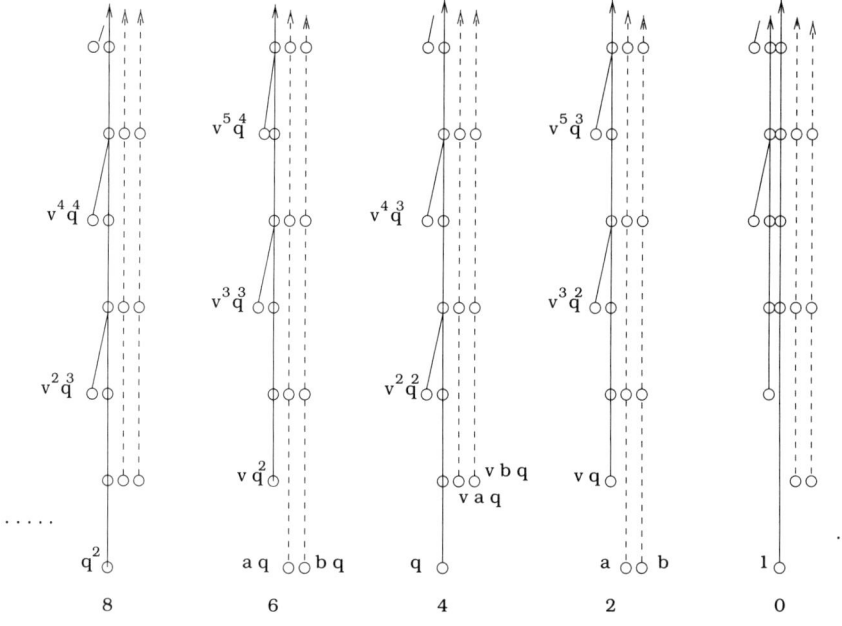

FIGURE 2.12. The Adams spectral sequence for $ku^*(BQ_8)$. (The action of v is omitted for clarity)

2.5. Dihedral groups.

The dihedral groups have the same representation rings as the quaternion groups of the same order, but have considerably more complicated H and ku-cohomologies, since they have rank 2. We let $D_{2^{n+2}} = \langle s, t | s^2 = t^2 = (st)^{2^{n+1}} = 1\rangle$.

The mod 2 cohomology ring is
$$H^*(BD_{2^{n+2}}; \mathbb{F}_2) = \mathbb{F}_2[x_1, x_2, w]/(x_2(x_1 + x_2))$$
with an $(n+1)$-st Bockstein $\beta_{n+1}(x_2w) = w^2$ ([**47**] and [**18**]). The generators are the Stiefel-Whitney classes of representions defined below:
$$\begin{aligned} x_1 &= w_1(\widehat{st}) \\ x_2 &= w_1(\widehat{s}) \\ w &= w_2(\sigma_1). \end{aligned}$$
By the Wu formula, $Sq^1(w) = x_1 w$ since $det(\sigma_1) = \widehat{st}$.

The representation ring is isomorphic to that of the quaternion group:
$$R(D_{2^{n+2}}) = \frac{\mathbb{Z}[\widehat{s}, \sigma_0, \sigma_1, \dots, \sigma_{2^n}]}{(\widehat{s}^2 = 1, \widehat{s}\sigma_i = \sigma_{2^n - i}, \sigma_i \sigma_j = \sigma_{i+j} + \sigma_{i-j})}$$
where we again let $\sigma_{-i} = \sigma_i$ and $\sigma_{2^n + i} = \sigma_{2^n - i}$. These are defined by
$$\sigma_i(s) = \begin{pmatrix} 0 & 1 \\ 1 & 0 \end{pmatrix} \quad \sigma_i(t) = \begin{pmatrix} 0 & \zeta^{-i} \\ \zeta^i & 0 \end{pmatrix}$$
$$\widehat{s}(s) = 1 \quad \widehat{s}(t) = -1$$
where ζ is a 2^{n+1}-st root of unity. Of these, all are irreducible except $\sigma_0 = 1 + \widehat{st}$ and $\sigma_{2^n} = \widehat{s} + \widehat{t}$, which are sums of one-dimensional representations pulled back from the abelianization $D_{2^{n+2}} \longrightarrow C_2 \times C_2 = <s, t \,|\, s^2 = t^2 = (st)^2 = 1>$. Again $\widehat{x}$ denotes the one-dimensional representation with kernel $<x>$.

The maximal subgroups are the kernels of these three nontrivial one-dimensional representations.

$$\begin{array}{ccc} & D_{2^{n+2}} & \\ \nearrow & \uparrow & \nwarrow \\ D_{2^{n+1}} & C_{2^{n+1}} & D_{2^{n+1}} \\ \| & \| & \| \\ <s, (st)^2> & <st> & <t, (st)^2> \end{array}$$

The difference from the quaternion groups shows up in the Euler classes. In particular, the determinant of σ_i is always $\widehat{st}$, independent of i, resulting in slightly different Euler classes in representation theory. The following ku-theory Euler classes do still generate $ku^*BD_{2^{n+2}}$:
$$\begin{aligned} a &= e_{ku}(\widehat{st}) &\in& \quad ku^2 BD_{2^{n+2}} \\ b &= e_{ku}(\widehat{s}) &\in& \quad ku^2 BD_{2^{n+2}} \\ d_i &= e_{ku}(\sigma_i) &\in& \quad ku^4 BD_{2^{n+2}}. \end{aligned}$$

As with the quaternion group, it will also be convenient to define
$$d = d_1 \in ku^4 BD_{2^{n+2}}$$
since we again do not need the other d_i to generate $ku^* BD_{2^{n+2}}$.

LEMMA 2.5.1. *For* $k = 1, \ldots, 2^n$,
$$d_k = \sum_{i=1}^{k} a_{ik} v^{2i-2} d^i$$
where $a_{kk} = (-1)^k$ *and* $a_{1k} = k^2$.

LEMMA 2.5.2. *The regular representation of* $D_{2^{n+2}}$,
$$\begin{aligned}\rho &= \sigma_0 + 2(\sigma_1 + \cdots + \sigma_{2^n-1}) + \sigma_{2^n} \\ &= v^{2^{n+1}} d^{2^n} + \left(\sum_{i=1}^{2^n-1} m_i v^{2i} d^i\right) + 2^{n+2} - 2^{n+1} va\end{aligned}$$
for integers m_i.

REMARK 2.5.3. The polynomials d_k are the Chebyshev polynomials conjugated by $F(d) = 1 - d/2$ ([**34**, Thm 4.4, p. 195]). This accounts for the remarkable composition property, $d_k(d_j) = d_{kj}$, and follows from the recursion formula
$$d_{k+1} = (2-d)d_k - d_{k-1} + 2d$$
with initial conditions $d_0 = 0$ and $d_1 = d$. ([**34**, (1.101), p.40]).

This is slightly simpler than for the quaternion groups, for which the corresponding Euler class q_k also involves a if $k \equiv 2 \pmod 4$. Correspondingly, the expression for the regular representation is slightly more complicated than for the quaternions.

THEOREM 2.5.4. $ku^*(BD_{2^{n+2}}) = (ku^*[a, b, d]/I)_J^\wedge$ *where* $|a| = |b| = 2$ *and* $|d| = 4$, *and where* I *is the ideal*
$$\begin{aligned}I = (&d\rho, va^2 - 2a, vb^2 - 2b, \\ &vad, 2ad, ab - b^2 + d_{2^n}, \\ &vbd - d + d_{2^n-1} - d_{2^n}, \\ &2bd - vdd_{2^n})\end{aligned}$$

The natural map $ku^* BD_{2^{n+2}} \longrightarrow H^* BD_{2^{n+2}}$ *sends* a *to* x_1^2, b *to* x_2^2, d_{2k} *to* 0, *and* d_{2k+1} *to* w^2.

For D_8, some of these relations simplify.

THEOREM 2.5.5. $ku^*(BD_8) = (ku^*[a, b, d]/I)_J^\wedge$ *where* $|a| = |b| = 2$ *and* $|d| = 4$, *and where* I *is the ideal*
$$\begin{aligned}I = (&v^4 d^3 - 6v^2 d^2 + 8d, \\ &va^2 - 2a, vb^2 - 2b, \\ &vad, 2ad, ab - b^2 + vbd, \\ &vbd - 4d + v^2 d^2, \\ &2bd - v^2 bd^2)\end{aligned}$$

Our proof of Theorems 2.5.4 and 2.5.5 will follow the same general outline as for the quaternion group. We first prove Lemma 2.5.2 and show that Lemma 2.5.1 holds after multiplication by v^2. We are then able to compute enough of the Adams

2.5. DIHEDRAL GROUPS.

spectral sequence to see that the $ku^*BD_{2^{n+2}}$ will be detected in mod 2 cohomology together with periodic K-theory. This allows us to then prove the theorems.

It will help to have the following relations between the ku-theory Euler classes and the representations in evidence:

$$\begin{aligned} va &= 2 - \sigma_0 \\ vb &= 1 - \widehat{s} \\ v^2 d &= \sigma_0 - \sigma_1 \\ v^2 d_i &= \sigma_0 - \sigma_i \end{aligned}$$

and therefore

$$\begin{aligned} \widehat{st} &= 1 - va \\ \widehat{s} &= 1 - vb \\ \widehat{t} &= (1-va)(1-vb) \\ \sigma_0 &= 2 - va \\ \sigma_i &= 2 - va - v^2 d_i \\ \sigma_{2^n} &= (2-va)(1-vb) \end{aligned}$$

Proof of Lemma 2.5.2: First, if we multiply the relation in Lemma 2.5.1 by v^2, we have a relation in the representation ring which follows inductively from the relation $\sigma_1 \sigma_k = \sigma_{k-1} + \sigma_{k+1}$ just as for the quaternion group (Lemma 2.4.3), but more easily, since the parity of k no longer has an effect. Then the expression for ρ in terms of the Euler classes is immediate. □

For brevity, let us write D for $D_{2^{n+2}}$ where there is no chance of confusion, and write H for $H\mathbb{F}_2$.

LEMMA 2.5.6. *In the Adams spectral sequence* $\mathrm{Ext}^{*,*}_{E(1)}(\mathbb{F}_2, H^*BD) \Longrightarrow ku^*BD$

(1) *there is one nonzero differential* d_{n+1},
(2) $E_{n+2} = E_\infty$ *is generated over* $\mathbb{F}_2[h_0, v]$ *by the filtration 0 classes* x_1^2, x_2^2, *and* w^2, *detecting* a, b, *and* d, *respectively, and*
(3) *multiplication by v is a monomorphism in positive Adams filtrations, and in codegrees less than or equal to 4.*

The natural map $ku^*BD \longrightarrow H^*BD \oplus K^*BD$ *is a monomorphism.*

Proof: Clearly, the final statement follows from (3). That a, b and d are detected by x_1^2, x_2^2 and w^2 is an immediate consequence of the naturality of Euler classes, since $x_1 = w_1(\widehat{st})$, $x_2 = w_1(\widehat{s})$ and $w = w_2(\sigma)$.

To compute the Adams spectral sequence we first decompose $H^*(BD)$ as an $E(1)$-module. Since the $E(1)$-module structure of $H^*(BD_{2^{n+2}})$ does not depend upon n (so long as n is greater than 0), we may use Bayen's ([5]) $\mathcal{A}(1)$ decomposition of H^*BD_8. From that, it is a simple matter to see that as an $E(1)$-module, H^*BD is the direct sum of

(1) two copies of $H^*(BC_2)$, $<x_1>$ and $<x_2>$,
(2) the trivial module $\{w^{2k+2}\}$, for each $k \geq 0$,
(3) free modules generated by

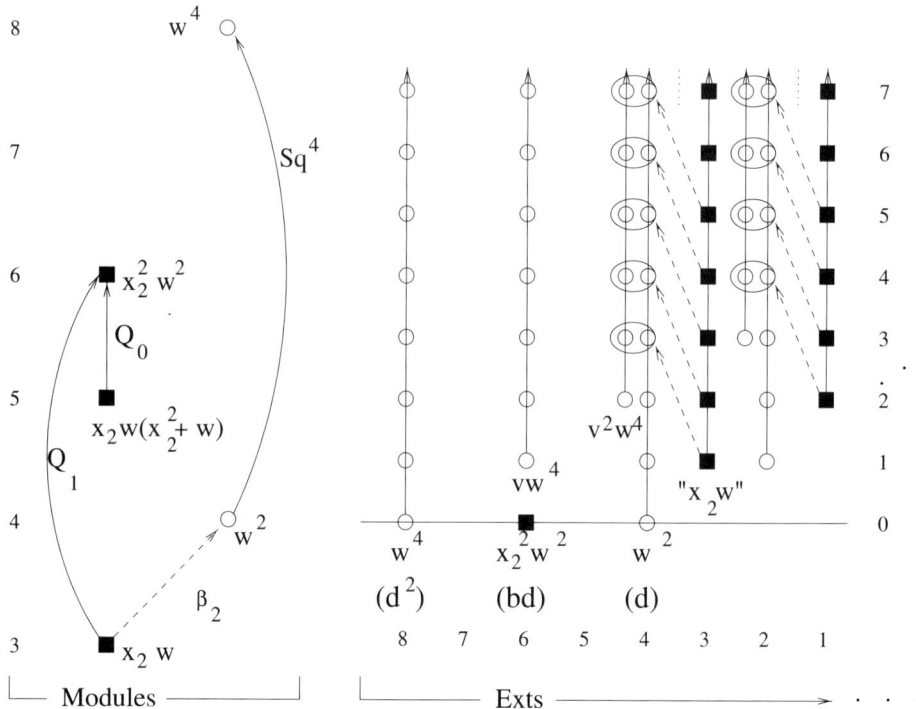

FIGURE 2.13. The beginning of summands (2) and (4) and their Ext modules for $D = D_8$

(a) $x_1^{2m} w^{2k+1}$ for each $m \geq 0$, $k \geq 0$, and
(b) $x_2^{2m} w^{2k+1}$, for each $m > 0$, $k \geq 0$, and
(4) the augmentation ideal, $IE(1) \cong \{x_2 w^{2k+1}, x_2 w^{2k+1}(x_2^2 + w), x_2^2 w^{2k+2}\}$ for each $k \geq 0$.

This allows us to write E_2 additively as the sum of

(1) $\mathbb{F}_2[h_0, v][a]/(va^2 - h_0 a) \oplus \mathbb{F}_2[h_0, v][b]/(vb^2 - h_0 b)$,
(2) $\mathbb{F}_2[h_0, v][d]$,
(3) an $\mathbb{F}_2$-vector space spanned by $\{a^i d^j | i \geq 1, j \geq 1\} \cup \{b^i d^j | i > 1, j \geq 1\}$
(4) the direct sum of the F_2-vector space spanned by $\{bd^j | j \geq 1\}$ and a free $\mathbb{F}_2[h_0, v]$-module on classes w^{2j} "$x_2 w$" $\in \text{Ext}^{1,4j-2}$. (We take the liberty of writing "$x_2 w$" for the nonzero element in $\text{Ext}^{1,-2}$ since $x_2 w$ is the element of mod 2 cohomology which corresponds to the tower generated by "$x_2 w$" in the theorem of May and Milgram. We will show that this class will support a differential and will therefore not occur in E_∞. Hence this sloppiness of notation should not cause any confusion. Strictly speaking, such a name would be justified only for an element of Ext^0, which is why we use quotation marks to warn that something is slightly amiss.)

Since a, b and d are Euler classes, they must survive to E_∞. This means that the only possible differentials originate from the $\mathbb{F}_2[h_0, v]$-free part of summand (4), and that the ring structure will determine all the differentials once we determine the first one.

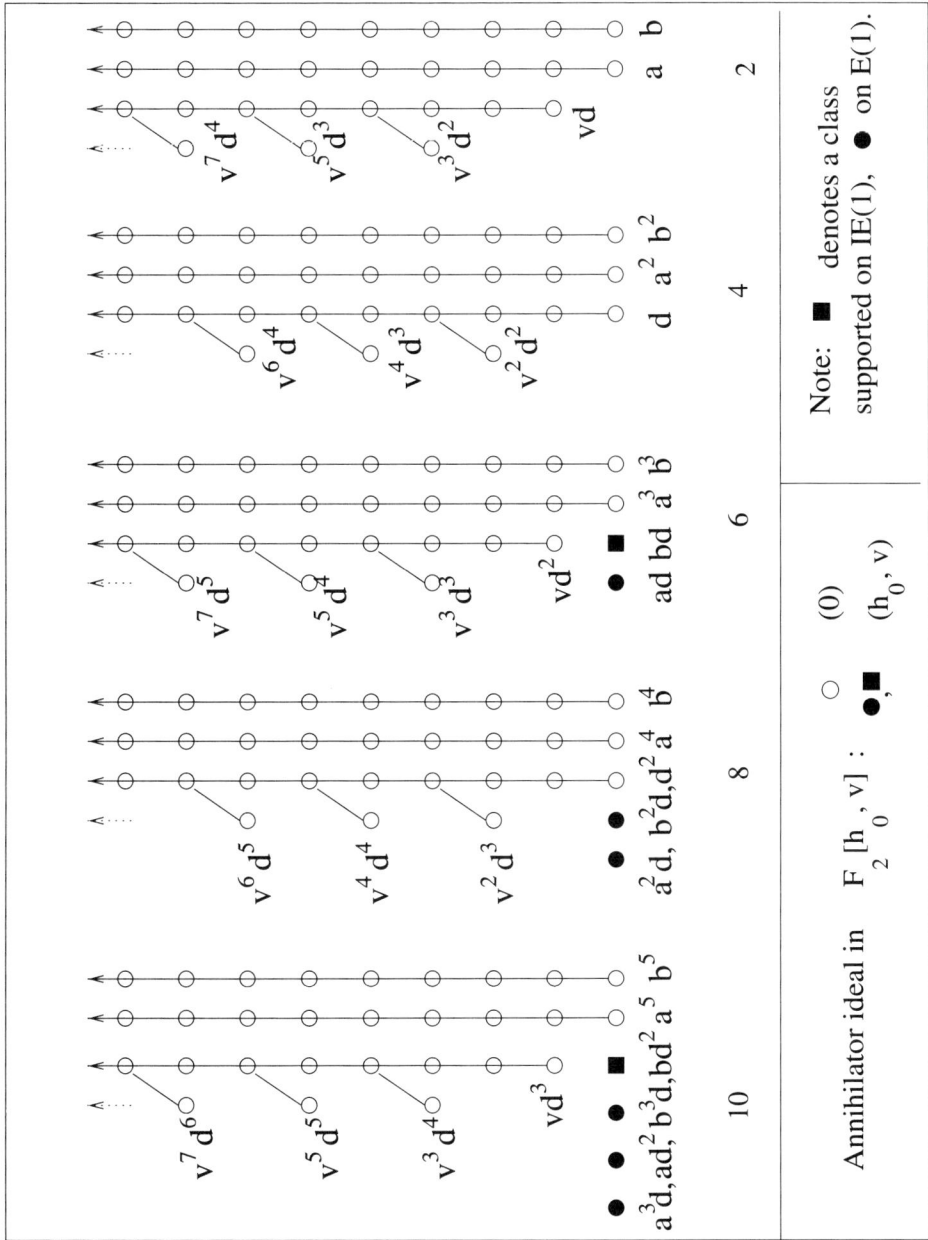

FIGURE 2.14. The E_∞ term of the Adams spectral sequence $\mathrm{Ext}_{E(1)}(\mathbb{F}_2, H^*(BD_8)) \Longrightarrow ku^*(BD_8)$.

In Figure 2.13 we show the beginning of summands (2) and (4) and their Ext modules for $D = D_8$. In the general case, replace β_2 by β_{n+1} and d_2 by d_{n+1}, and increase the number of towers in the even stems appropriately.

The Bockstein $\beta_{n+1}(x_2 w) = w^2$ implies that there will be an Adams d_{n+1} which connects towers in these codegrees. It must take the form

$$d_{n+1}(\text{``}x_2w\text{''}) = a_0 h_0^{n+2} w^2 + vx$$

for some $a_0 \in \mathbb{F}_2$ and some x. Restricting to the 4-skeleton shows that $a_0 = 1$. Since $h_0^{n+2} w^2 + vx$ is not a zero-divisor in positive filtration, E_{n+2} is concentrated in even codegrees, and is therefore equal to E_∞. Further, since v does not divide $h_0 w^2$, v acts monomorphically on positive filtrations. Inspection of the E_2 term shows this also holds for filtration 0 in codegrees 4 and below. □

We can now finish the calculation. First, we show that the stated relations hold, and then that no others are needed.

Proof of Lemma 2.5.1: In the proof of Lemma 2.5.2 we showed that the formula for d_k holds after multiplication by v^2, and we have just shown that this is a monomorphism. □

LEMMA 2.5.7. *The relations in Theorem 2.5.4 hold in $ku^* BD_{2^{n+2}}$.*

Proof: These are straightforward. Since ρ is induced up from the trivial subgroup, $d\rho = 0$. Since $\widehat{s}^2 = 1$, $vb^2 = 2b$. Similarly, $\sigma_0^2 = 2\sigma_0$ implies that $va^2 = 2a$.

Now, $\sigma_0 \sigma_i = 2\sigma_i$ implies that $v^3 a d_i = 0$ and Lemma 2.5.6.3 implies that $v a d_i$ is therefore 0. Hence, $2 a d_i = va^2 d_i = 0$ as well. In particular, $vad = 0$ and $2ad = 0$.

To see that $ab = b^2 - d_{2^n}$ we use Lemma 2.5.6 and test the relation in both cohomology and representation theory. In cohomology it follows from $x_1 x_2 = x_2^2$ and in representation theory it follows from $\widehat{s}\sigma_0 = \sigma_{2^n}$

The relation $vbd = d - d_{2^n - 1} + d_{2^n}$ follows from $v^3 bd = (1 - \widehat{s})(\sigma_0 - \sigma_1)$ in the representation ring, since it is in positive Adams filtration.

Finally, $2bd = vdd_{2^n}$ also need only be checked in the representation ring. There, it says $2(1 - \widehat{s})(\sigma_0 - \sigma_1) = (\sigma_0 - \sigma_{2^n})(\sigma_0 - \sigma_1)$, which is easily verified. □

Now, the Adams spectral sequence (Lemma 2.5.6) tells us that in positive codegrees the elements

$$\{a^i d^k, b^i d^k | i \geq 0, k \geq 0\} \cup \{v^i d^k | i > 0, k \geq 0\}$$

generate $ku^* BD$ as a $\mathbb{Z}_2^\wedge$-module. It remains to determine the relations among these, and show they all follow from the relations already found. We shall do this using the monomorphism from $ku^* BD$ into $H^* BD \oplus K^* BD$. Let us give names to the images of a, b, d, and d_k in the representation ring $R(D)$:

$$\begin{aligned} A &= 1 - \widehat{st} \\ B &= 1 - \widehat{s} \\ D &= \sigma_0 - \sigma_1 \\ D_k &= \sigma_0 - \sigma_k \end{aligned}$$

and note that $A^2 = 2A$, $B^2 = 2B$, and $2BD = DD_{2^n}$. The situation in codegrees $4k$ and $4k + 2$ is slightly different, so we consider them separately. In codegree

$4k+4$ our generators and their images in H^*BD and K^*BD are:

$$
\begin{array}{llll}
a^{2k+2} & x_1^{4k+4} & 2^{2k+1}A & \\
b^{2k+2} & x_2^{4k+4} & 2^{2k+1}B & \\
d^k d_i & iw^{2(k+i)} & D^k D_i & i = 1, \ldots, 2^n \\
a^{2i+2}d^{k-i} & x_1^{4i+4}w^{2(k-i)} & 0 & i = 0, \ldots, k-1 \\
b^{2i+2}d^{k-i} & x_2^{4i+4}w^{2(k-i)} & 2^{2i+1}BD^{k-i} & i = 0, \ldots, k-1.
\end{array}
$$

In codegree $4k+2$ we have:

$$
\begin{array}{llll}
a^{2k+1} & x_1^{4k+2} & 2^{2k}A & \\
b^{2k+1} & x_2^{4k+2} & 2^{2k}B & \\
vd^k d_i & 0 & D^k D_i & i = 1, \ldots, 2^n \\
a^{2i+1}d^{k-i} & x_1^{4i+2}w^{2(k-i)} & 0 & i = 0, \ldots, k-1 \\
b^{2i+1}d^{k-i} & x_2^{4i+2}w^{2(k-i)} & 2^{2i}BD^{k-i} & i = 0, \ldots, k-1.
\end{array}
$$

The generators a^i, b^i, and $v^\epsilon d^k d_i$ span a copy of $(\mathbb{Z}_2^\wedge)^{2^n+2}$ since their images in K^*BD do. The generators $a^i d^j$ generate an $\mathbb{F}_2 = ku^*/(2,v)$ vector space since their image in K^*BD is zero. The only complications come from the relations between $b^i d^j$ and the other generators. Since $2bd = vdd_{2^n}$, we see that the generator vdd_{2^n} is redundant, and should be replaced by bd. In general, the image of $b^{2i+1}d^{k-i}$ in K^*BD is linearly dependent upon the images of the $vd^k d_i$, and similarly in codegrees $4k+4$. To make this explicit, when $0 < i < k$ we find that

$$
\begin{aligned}
B^{2i+1}D^{k-i} &= 2^{2i}BD^{k-i} \\
&= 2^{2i-1}(2BD)D^{k-i-1} \\
&= 2^{2i-1}(DD_{2^n})D^{k-i-1} \\
&= 2^{2i-1}D^{k-i}D_{2^n} \\
&= D^k a_i \cdot (D_1, \ldots, D_{2^n})
\end{aligned}
$$

if $a_i \cdot (D_1, \ldots, D_{2^n})$ solves $D^i a_i \cdot (D_1, \ldots, D_{2^n}) = 2^{2i-1}D_{2^n}$. If we then let $\alpha_i = a_i \cdot (d_1, \ldots, d_{2^n})$, we find that $b^{2i+1}d^{k-i}$ and $vd^k \alpha_i$ have the same image in K^*BD but not in H^*BD, so their difference is annihilated by 2 and v. In codegrees $4k+4$, the situation is nearly the same: $b^{2i+2}d^{k-i}$ and $2d^k \alpha_i$ have the same image in K^*BD but not in H^*BD, so their difference is annihilated by 2 and v. In both codegrees, the case $i = 0$ is slightly different: since $2bd^k = vd^{k-1}d_{2^n}$ no torsion class is generated. Similarly, $b^2 d^k$ maps to $D^k D_{2^n}$ in K^*BD, so $d^k d_{2^n}$ replaces the nonexistent $2d^k \alpha_0$.

We will thus have a complete understanding of the additive structure once we show that the equations $D^i a_i \cdot (D_1, \ldots, D_{2^n}) = 2^{2i-1}D_{2^n}$ can be solved. To this end, consider the summand $\langle D_1, \ldots, D_{2^n} \rangle$ in $R(D)$.

LEMMA 2.5.8. *In the representation ring $R(D)$,*

(1) *multiplication by* $D = D_1$ *sends* $\langle D_1, \ldots, D_{2^n} \rangle$ *to itself by the matrix*

$$\Delta = \begin{pmatrix} 4 & 1 & 2 & 2 & \cdots & 2 & 2 \\ -1 & 2 & -1 & 0 & & 0 & 0 \\ & -1 & 2 & -1 & & \vdots & \vdots \\ & & -1 & 2 & & 0 & 0 \\ & & & -1 & \ddots & -1 & 0 \\ & & & & & 2 & -2 \\ & & & & & -1 & 2 \end{pmatrix}$$

(2) *For each* $i > 0$, *there exists an integer vector* a_i *such that* $\sum a_{ij} D_j = a_i \cdot (D_1, \ldots, D_{2^n})$ *satisfies* $\Delta^i a_i \cdot (D_1, \ldots, D_{2^n}) = 2^{2i-1} D_{2^n}$.

Proof: Part (1) follows from the recursion relation

$$D_{i+1} = (2 - D)D_i - D_{i-1} + 2D$$

which says

$$DD_i = 2D - D_{i-1} + 2D_i - D_{i+1}$$

for $i < 2^n$. When $i = 2^n$ this becomes

$$\begin{aligned} DD_{2^n} &= 2D - D_{2^n-1} + 2D_{2^n} - D_{2^n+1} \\ &= 2D - 2D_{2^n-1} + 2D_{2^n} \end{aligned}$$

using $D_{2^n+1} = D_{2^n-1}$, and this follows from $d\rho = 0$, in the form

$$0 = 2^{n+2}d - 2d(d_1 + \cdots + d_{2^n-1}) - dd_{2^n}$$

by the preceding relations. This determines the matrix Δ.

For part (2) the vector a' with i-th coordinate $i - 2^{n-1}$ for $1 \leq i < 2^n$ and with 2^n-th coordinate 2^{n-2} satisfies $\Delta a' = e_{2^n} = (0, 0, \ldots, 0, 1)$. Thus, $a_1 = 2a'$ is the solution when $i = 1$. If we suppose inductively that a_i solves $\Delta a_i = 2^{2i-1} e_{2^n}$ and has $a_{ij} + a_{i,2^n-j} = 0$, it is simple to check that we can find an integer vector a_{i+1} satisfying $\Delta a_{i+1} = 4a_i$ and that it has the same symmetry property. $\square$

If we now write

$$\alpha_i = a_i \cdot (d_1, \ldots, d_{2^n}) \in ku^4 BD$$

for each $i \geq 1$, we can neatly express the ku^*-module structure of $ku^* BD$.

LEMMA 2.5.9. (1) *If* $k \leq 0$ *then* $ku^{2k}(BD) = \mathbb{Z} \oplus (\mathbb{Z}_2^\wedge)^{2^n+2}$, *generated by*

$$v^k \{1, va, vb, v^2 d, v^2 d_2, \ldots, v^2 d_{2^n}\}.$$

(2) $ku^2(BD) = (\mathbb{Z}_2^\wedge)^{2^n+2}$ *generated by* $\{a, b, vd, vd_2, \ldots, vd_{2^n}\}$.
(3) $ku^4(BD) = (\mathbb{Z}_2^\wedge)^{2^n+2}$ *generated by* $\{a^2, b^2, d, d_2, \ldots, d_{2^n}\}$.
(4) *If* $k > 0$ *then* $ku^{4k+2}(BD) = (\mathbb{Z}_2^\wedge)^{2^n+2} \oplus \mathbb{F}_2^{2k-1}$ *with* $(\mathbb{Z}_2^\wedge)^{2^n+2}$ *generated by*

$$\{a^{2k+1}, b^{2k+1}, vd^k d_1, \ldots, vd^k d_{2^n-1}, bd^k\},$$

an $\mathbb{F}_2^k$ *generated by*

$$\{ad^k, a^3 d^{k-1}, \ldots, a^{2k-1} d\},$$

and an $\mathbb{F}_2^{k-1}$ *generated by*

$$\{b^3 d^{k-1} - vd^k \alpha_1, \ldots, b^{2k-1} d - vd^k \alpha_{k-1}\}.$$

(5) If $k > 0$ then $ku^{4k+4}(BD) = (\mathbb{Z}_2^{\wedge})^{2^n+2} \oplus \mathbb{F}_2^{2k}$ with $(\mathbb{Z}_2^{\wedge})^{2^n+2}$ generated by
$$\{a^{2k+2}, b^{2k+2}, d^k d_1, d^k d_2, \ldots, d^k d_{2^n}\},$$
an $\mathbb{F}_2^k$ generated by
$$\{a^2 d^k, a^4 d^{k-1}, \ldots, a^{2k} d\},$$
and an $\mathbb{F}_2^k$ generated by
$$\{abd^k = (b^2 - d_{2^n})d^k, b^4 d^{k-1} - 2d^k \alpha_1, \ldots, b^{2k}d - 2d^k \alpha_{k-1}\}.$$

(6) The 2-torsion is annihilated by v.

Proof: The Adams spectral sequence determines the structure of $ku^i BD$ for $i \leq 4$, as there is no torsion in these degrees, and the map into $K^* BD$ is a monomorphism. This establishes (1)-(3). To show (6) simply observe that if vx is nonzero then x maps nontrivially to $K^* BD$, so cannot have finite additive order.

Before the lemma, we showed that parts (4) and (5) would follow from the lemma. □

Proof of Theorem 2.5.4: It is now a simple matter to verify that the stated relations are complete, by checking that multiplying any of the additive generators of the preceding lemma by a, b, d, or v produces an element which can be written in terms of that additive basis using the relations. □

REMARK 2.5.10. Some interesting facts emerge from these relations.

(1) There is a large filtration shift in the relation $2bd = vdd_{2^n}$. The class bd is in Adams filtration 0, while 2 times it is in Adams filtration $2n+1$.

(2) The relations $q\rho = 0$ and $d\rho = 0$ for the quaternion and dihedral groups are exactly the same polynomial in q and d respectively. This implies that 2^{n+2} will annihilate their images in cohomology. For $Q_{2^{n+2}}$ this is exactly the order of q, but for $D_{2^{n+2}}$, the additional relation $vbd = d + d_{2^n} - d_{2^n-1}$ makes $2^{n+1}d$ divisible by v, so that the image of d in cohomology therefore has order 2^{n+1}. This is an example of the relation between the v-filtration of the representation ring and cohomology.

2.6. The alternating group of degree 4.

The ku-cohomology of the group A_4 has two interesting features. First, the Euler class of the 3-dimensional irreducible representation is nonzero in connective K-theory, but is 0 in periodic K-theory. Second, $ku^*(BA_4)$ is not generated by Chern classes. (Neither is $H^*(BA_4)$.)

The representation ring is
$$R(A_4) = \mathbb{Z}[\alpha, \tau]/(\alpha^3 - 1, \tau(\alpha - 1), \tau(\tau - 2) - 1 - \alpha - \alpha^2)$$
where α factors through the quotient $A_4 \longrightarrow C_3$, and τ is the reduced permutation representation. We define
$$\begin{aligned} A &= e_K(\alpha) = (1-\alpha)/v \\ \Gamma &= (1 + \alpha + \alpha^2 - \tau)/v^3 \end{aligned}$$
and note that the augmentation ideal $J = (vA, v^3 \Gamma)$. This gives us the periodic K-theory.

PROPOSITION 2.6.1. $K^*(BA_4) = (K^*[A, \Gamma]/I)^\wedge_j$ where I is the ideal
$$I = ((1-vA)^3 - 1,\ A\Gamma,\ \Gamma(v^3\Gamma - 4)).$$

Additively, we have $K^0(BA_4) = \mathbb{Z} \oplus (\mathbb{Z}_3^\wedge)^2 \oplus \mathbb{Z}_2^\wedge$, generated by 1, vA, v^2A^2, and $v^3\Gamma$, respectively. To compute the connective K-theory, consider the extension

$$V_4 \xrightarrow{\triangleleft} A_4 \longrightarrow C_3.$$

Note that we are using the traditional notation V_4 for the Klein 4-group here. That is, V_4 has *order* 4, and not *rank* 4. The Atiyah-Hirzebruch spectral sequences collapse, giving

$$H^*(BA_4)_{(3)} = H^*(BC_3)_{(3)}, \qquad ku^*(BA_4)_{(3)} = ku^*(BC_3)_{(3)}$$
$$H^*(BA_4)_{(2)} = H^*(BV_4)^{C_3}_{(2)}, \qquad ku^*(BA_4)_{(2)} = ku^*(BV_4)^{C_3}_{(2)}$$

First, consider the 2-localization of $ku^*(BA_4)$. Let $H = H\mathbb{F}_2$. Then $H^*(BV_4) = \mathbb{F}_2[x_i, x_j]$, with $x_k = x_i + x_j$ and C_3-action $x_i \mapsto x_j \mapsto x_k$. The resulting ring of invariants is
$$H^*(BA_4; \mathbb{F}_2) = \mathbb{F}_2[a, b, c]/(a^3 + b^2 + bc + c^2)$$
with $|a| = 2$, and $|b| = |c| = 3$. The inclusion into $H^*(BV_4)$ sends
$$a \mapsto x_i^2 + x_i x_j + x_j^2,$$
$$b \mapsto x_i x_j (x_i + x_j),$$
$$c \mapsto x_i^3 + x_i^2 x_j + x_j^3.$$

We will see in section 4.2 that $ku^*(BV_4) = (ku^*[y_i, y_j]/(vy_i^2 - 2y_i, vy_j^2 - 2y_j))^\wedge_j$. The C_3-action is $y_i \mapsto y_j \mapsto y_k = y_i + y_j - vy_iy_j$. The resulting ring of invariants is what we must compute. The Chern classes of τ will give two of the three generators. They are
$$\nu = e_{ku}(\tau) = c_3(\tau) \quad \in ku^6(BA_4)$$
$$\mu = \qquad\qquad c_2(\tau) \quad \in ku^4(BA_4)$$
$$v\mu = \qquad\qquad c_1(\tau) \quad \in ku^2(BA_4)$$

The remaining generator will be defined by its image in $ku^*(BV_4)$.

THEOREM 2.6.2. $ku^*(BA_4)_{(2)} = (ku^*[\mu, \nu, \pi]/I)^\wedge_j$ where $|\mu| = 4$, $|\nu| = |\pi| = 6$, and where I is the ideal
$$I = (2\nu,\ v\nu,\ v\pi - 2\mu,\ v\mu^2 - 2\pi,\ \mu^3 - \pi^2 - \pi\nu - \nu^2)$$

The natural map $ku^*(BA_4) \longrightarrow K^*(BA_4)$ sends ν to 0, μ to $v\Gamma$, and π to 2Γ. The natural map $ku^*(BA_4) \longrightarrow H\mathbb{F}_2^*(BA_4)$ sends μ to a^2, ν to b^2, and π to c^2. The restriction to $ku^*(BV_4)$ sends
$$\mu \mapsto y_i^2 - y_i y_j + y_j^2,$$
$$\nu \mapsto y_i y_j (y_i - y_j),$$
$$\pi \mapsto y_i^3 - y_i^2 y_j + y_j^3.$$

Proof: Since $ku^*(BA_4)_{(2)}$ consists of the C_3-invariants in $ku^*(BV_4)$, we could use the images of μ, ν and π to define them. It is then a simple calculation with Chern classes in $ku^*(BV_4)$ to verify that this agrees with the definitions of μ and ν already given. To see that these generate the invariants and that we have all the relations, we shall use the Adams spectral sequence.

Recall that there is a stable 2-local splitting
$$ku \wedge BA_4 = (ku \wedge \Sigma^2 BC_2) \vee \text{GEM},$$

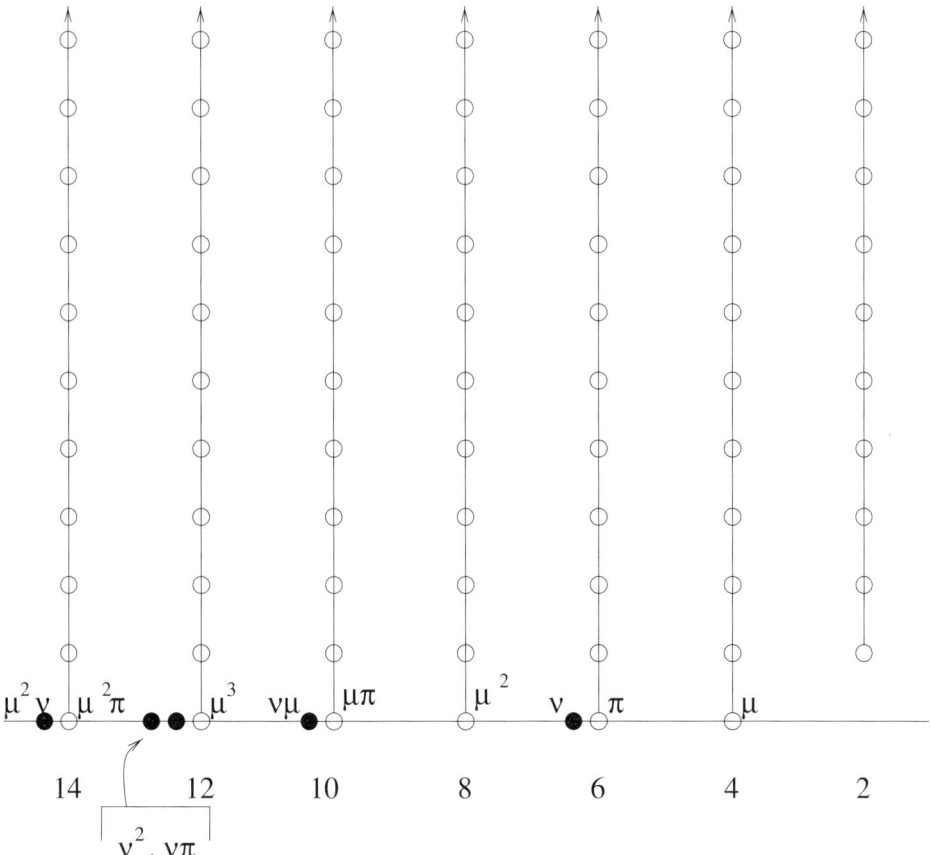

FIGURE 2.15. The $E_2 = E_\infty$ term of the Adams spectral sequence for $ku^*(BA_4)$.

where the GEM is a wedge of $H\mathbb{F}_2$'s. This follows from the splitting

$$BC_2 \wedge BC_2 = BA_4 \vee \bigvee_\alpha \Sigma^{n_\alpha} L(2)$$

and the equivalence

$$ku \wedge BC_2 \wedge BC_2 = ku \wedge \Sigma^2 BC_2 \vee \Sigma^2 H\mathbb{F}_2[u,v]$$

with $|u| = |v| = 2$. This implies that the Adams spectral sequence for $ku^*(BA_4)_{(2)}$ collapses at E_2, since this is true for both $ku \wedge \Sigma^2 BC_2$ and $H\mathbb{F}_2$. (See Figure 2.15.) It also implies that additive generators for $ku^*(BA_4)_{(2)}$ are all detected in mod 2 cohomology. Now, the elements μ, ν, and π are detected by a^2, b^2, and c^2 in $H\mathbb{F}_2^*(BA_4)$, and a check of Poincaré series shows that the algebra generated by these three elements accounts for Ext^0, which consists of the (Q_0, Q_1)-annihilated elements of $H\mathbb{F}_2^*(BA_4)$. Thus, these three elements will generate $ku^*(BA_4)_{(2)}$. The relations follow immediately from the relations in $ku^*(BV_4)$, and the images in $K^*(BA_4)$ follow by restriction to V_4 as well. □

Now we can assemble the 2-local and 3-local information.

THEOREM 2.6.3. $ku^*(BA_4) = (ku^*[y, \mu, \nu, \pi]/I)^\wedge_J$ where $|y| = 2$, $|\mu| = 4$, $|\nu| = |\pi| = 6$, and where I is the ideal

$$\begin{aligned} I \;=\; & ([3](y), y\mu, y\nu, y\pi, \\ & 2\nu, v\nu, v\pi - 2\mu, v\mu^2 - 2\pi, \mu^3 - \pi^2 - \pi\nu - \nu^2). \end{aligned}$$

Proof: The integral result injects into the sum of the localizations, so the only question is the relations between y and μ, ν and π. But $3y \in (v)y$, where (v) is the ideal generated by v, while $4\mu \in (v)\mu$, $2\nu = 0$, and $4\pi \in (v)\pi$. Since the v-adic filtration is complete, the products $y\mu$, $y\nu$, and $y\pi$ must all be 0. $\square$

CHAPTER 3

The ku-homology of finite groups.

The purpose of this chapter is to explain and illustrate how to use the cohomology calculations to deduce their homology and Tate cohomology, and to discuss the associated duality. For some time it has been routine to obtain additive information using the Adams spectral sequence methods, at least up to extension [**5, 6**]. Indeed, our calculations of the cohomology used some of this information. However we are using the local cohomology spectral sequence to deduce $ku_*(BG)$ as a module over $ku^*(BG)$ from a knowledge of the ring $ku^*(BG)$ and the Euler classes. This has purely practical advantages in the ease of calculating certain additive extensions, but the main motivation is to obtain structural and geometric information not accessible through the Adams spectral sequence. In particular we expose the remarkable duality properties of $ku^*(BG)$ which follow from the local cohomology theorem.

We begin by giving a more detailed discussion of generalities on the use of the local cohomology theorem in Section 3.1, the duality in Section 3.3 and Tate cohomology in Section 3.6. In each section, we end with specific examples.

3.1. General behaviour of $ku_*(BG)$.

One reason for understanding $ku^*(BG)$ as a ring is the local cohomology theorem which states that there is a spectral sequence
$$H_I^{*,*}(ku^*(BG)) \Rightarrow ku_*(BG)$$
where $I = ker(ku^*(BG) \longrightarrow ku^*)$ is the augmentation ideal and $H_I^*(\cdot)$ denotes the local cohomology functor. Local cohomology can be calculated using a stable Koszul complex, and it calculates the right derived functors of the I-power torsion functor
$$\Gamma_I(M) = \{m \in M \mid I^s m = 0 \text{ for } s >> 0\}$$
on ku^*-modules M. This is proved as in [**22**, Appendix] and spelled out in [**31**]. What is required is an equivariant S-algebra $\widehat{ku}$ with Noetherian coefficient ring $ku^*(BG)$. A suitable S-algebra is $\widehat{ku} = F(EG_+, \inf_1^G ku)$, where $\inf_1^G$ is the highly structured inflation of Elmendorf-May [**17**]. If G is a p-group, the augmentation ideal I may be replaced by the ideal $\mathcal{E}(G)$ generated by Euler classes, or even $\mathcal{E}'(G) = (e_{ku}(V_1), e_{ku}(V_2), \ldots, e_{ku}(V_r))$ provided G acts freely on the product $S(V_1) \times S(V_2) \times \cdots \times S(V_r)$ of unit spheres [**23**]. The method of proof is described in [**22**, Section 3].

The analogue for periodic K-theory [**22**] is particularly illuminating because the representation ring is 1-dimensional, so the spectral sequence collapses to show $K_0(BG) = H_J^0(R(G)) = \mathbb{Z}$ and $K_1(BG) = H_J^1(R(G))$. The analogue for ordinary cohomology has been illuminating for different reasons. Many cohomology rings of finite groups are Cohen-Macaulay or have depth one less than their dimension.

In these cases the spectral sequence collapses, and this shows that the cohomology ring is actually Gorenstein or almost Gorenstein [29].

The connective K-theory ring $ku^*(BG)$ is more complicated. It is of I-depth 0 if G is non-trivial, and it is of dimension equal to the rank of G. Accordingly the spectral sequence does not often collapse. Nonetheless we have found this method of calculating $ku_*(BG)$ useful, and we illustrate the calculations in a number of cases including cyclic groups, generalized quaternion groups, the dihedral group of order 8, and most substantially, in Chapter 4, elementary abelian 2-groups of arbitrary rank. For the remainder of this section we describe the approximate form of our calculations.

For brevity we let $R = ku^*(BG)$, and note that $R[1/v] = K^*(BG)$ by Lemma 1.1.1. For the remainder of the discussion, suppose that G is a p-group. It is often useful to calculate local cohomology using the diagram

$$\begin{array}{ccccc} R & \longrightarrow & R[1/p] & \longrightarrow & R/p^\infty \\ \downarrow & & \downarrow & & \downarrow \\ R[1/v] & \longrightarrow & R[1/v][1/p] & \longrightarrow & R[1/v]/p^\infty. \end{array}$$

The principal tool is the first row, but note that the second row is the chromatic complex for periodic K-theory and therefore completely understood. Indeed it gives the exact sequence

$$0 \longrightarrow H_I^0(K^*(BG)) \longrightarrow K^*(BG)[1/p] \cdot \rho \longrightarrow K^*(BG)/p^\infty \longrightarrow H_I^1(K^*(BG)) \longrightarrow 0,$$

so that

$$H_I^0(K^*(BG)) = \mathbb{Z}[v, v^{-1}] \cdot \rho$$

and

$$H_I^1(K^*(BG)) = \{(R(G)/(\rho))/p^\infty\}[v, v^{-1}].$$

The comparison between the rows gives essential information. The first row is not so effective as in the periodic case, since there is often p-torsion in R. There are cases of interest when R has no p-torsion, but in general we obtain the two short exact sequences

$$\Gamma_p R \longrightarrow R \longrightarrow \overline{R} \quad \text{and} \quad \overline{R} \longrightarrow R[1/p] \longrightarrow R/p^\infty.$$

The p-power torsion $\Gamma_p R$ is often essentially on the 0-line of the Adams spectral sequence and therefore a submodule of the mod p cohomology of BG. Let d denote the depth of $H^*(BG; \mathbb{F}_p)$: typically this is fairly large, and in any case Duflot's theorem [13] states the depth is bounded below by the p-rank of the centre. We will speak as if this is also the depth of $\Gamma_p R$, although this is typically only true in an attenuated sense (see Chapter 4 for further discussion). Furthermore the local cohomology of $\Gamma_p R$ should be very well behaved in the top degrees (Gorenstein or almost Gorenstein). This says the local cohomology in most degrees (in degrees $\leq d-2$) is the same as that of $\overline{R}$. Now $H_I^*(R[1/p])$ is concentrated in local cohomology degree 0 (see 3.1.4 below) and therefore the local cohomology of $\overline{R}$ is that of R/p^∞ in one degree lower. In low rank examples this reduces the calculation of $H_I^*(R)$ to calculations of H_I^0, which can be regarded as routine. In general we have $H_I^s(R) = H_I^{s-1}(R/p^\infty)$ for $2 \leq s \leq d-2$, and there is also a tractable exact sequence for $H_I^0(R)$ and $H_I^1(R)$ in terms of I-power torsion.

In the two most difficult cases we have calculated (the elementary abelian and dihedral cases), these generalities have been usefully packaged as follows. We consider the short exact sequence
$$0 \longrightarrow T \longrightarrow R \longrightarrow Q \longrightarrow 0$$
of R-modules where T is the v-power torsion. Thus Q is the image of $ku^*(BG)$ in $K^*(BG)$, and in all cases except A_4 it is the modified Rees ring: the $K^0(BG)$ subalgebra of $K^*(BG)$ generated by $1, v$ and the K theory Chern classes of representations. Although T is defined as the v-power torsion, in our examples when G is a p-group it turns out to be the (p, v)-power torsion. In many cases it is annihilated by the exponent of the group without the need for higher powers.

In the examples, I-local cohomology of Q can be calculated as (y^*)-local cohomology for a suitable element $y^* \in I$. In the abelian case the reason is that, viewed in $R(G)$, the ideal (y^*) is a reduction of the augmentation ideal J, and the modification in general takes account of other Chern classes. This means that the local cohomology of Q is zero except in degrees 0 and 1. Since $Q \subseteq K^*(BG)$, it is easy to see that $H_I^0(Q) = ku^* \cdot \rho$. It can be seen that $H_I^1(Q)$ is $\mathbb{Z}$-torsion, and the order of the torsion increases with degree.

SUMMARY 3.1.1. *Provided T is of I-depth ≥ 2,*
$$H_I^0(R) = ku^* \cdot \rho,$$
there is an exact sequence
$$0 \longrightarrow H_I^1(R) \longrightarrow H_I^1(Q) \longrightarrow H_I^2(T) \longrightarrow H_I^2(R) \longrightarrow 0$$
and isomorphisms
$$H_I^i(R) \cong H_I^i(T)$$
for $i \geq 3$.

We believe the hypothesis in the summary is always satisfied.

CONJECTURE 3.1.2. *The I-depth of the v-power torsion submodule T is always ≥ 2.*

Before turning to examples we prove some general facts.

LEMMA 3.1.3. *(i) The $\mathbb{Z}$-torsion submodule is v-power torsion. (ii) The v-power torsion submodule T is $\mathbb{Z}$-torsion.*

PROOF. Part (i) is obvious since $K^*(BG)$ is $\mathbb{Z}$-torsion free.

For Part (ii) we need only note that $[X, K/ku] \otimes \mathbb{Q} = [X, K/ku \otimes \mathbb{Q}]$ for any bounded below spectrum X, because the limits in the Milnor exact sequence are eventually constant, and that $X = BG$ is rationally trivial. □

LEMMA 3.1.4. *The module $R[1/p]$ has local cohomological dimension 0 in the sense that $H_I^s(R[1/p]) = 0$ for $s > 0$.*

Proof: Method 1: Local cohomological dimension is detected on varieties, so by Quillen's descent argument (see Section 1.1) it is sufficient to check for abelian groups A. By the Künneth theorem up to varieties (1.5.1) it is enough to check on cyclic groups. This is true.

Method 2: Once the regular representation is factored out we have a vector space over $\mathbb{Q}_p$. The action by the Euler class e of a one dimensional representation has only the regular representation in the kernel. Thus e is an injective map of finite dimensional vector spaces (on $R[1/p]/\Gamma_I R[1/p]$) over $\mathbb{Q}_p$ and hence an isomorphism. $\square$

Since $ku_*(BG)$ is I-power torsion by the local cohomology theorem, the image of the norm map lies in $H^0_I(ku^*(BG))$. These two submodules are often equal.

LEMMA 3.1.5. *For a finite group G, we have $H^0_I(ku^*(BG)) = ku^* \cdot \rho$ provided either*

(1) *v is a regular element in $ku^*(BG)$, or*
(2) *the v-torsion of the p-Sylow subgroups of G is detected on the 0-line for all p.*

Proof: By 3.6.1 $ku^* \cdot \rho$ lies in the 0-th local cohomology. This gives a lower bound.

We also know by character theory that $H^0_I(K^*(BG)) = K^* \cdot \rho$. Therefore elements of $ku^*(BG)$ detected in $K^*(BG)$ are only I-power torsion if they are in $ku^* \cdot \rho$. If v acts monomorphically on $ku^*(BG)$, all elements of $ku^*(BG)$ are detected in $K^*(BG)$ and we are done.

In the other case, the fact that $ku^*(BG) = K^*(BG)$ in positive degrees shows that the answer is correct in positive degrees. The v-torsion is $\mathbb{Z}$-torsion, so it suffices to check for each p that there is no more p-local I-torsion. By a transfer argument, $ku^*(BG)_{(p)}$ is a retract of $ku^*(BP)_{(p)}$ for a P-Sylow subgroup P. We may therefore assume G is a p-group.

Thus it is enough to consider what happens in negative degrees on v-power torsion. By assumption, all v-power torsion is on the 0-line of the Adams spectral sequence. It suffices to find an element of the augmentation ideal which acts regularly on the 0-line. Now [14] shows there is an element of mod p cohomology with this property, and we show that it lifts to $ku^*(BG)$. Indeed, if p^n is the order of the largest abelian subgroup of G, for each abelian subgroup A, we may consider the total Chern class $c^{ku}_\cdot((p^n/|A|)\rho_A)$. Since $\text{res}^A_B \rho_A = (|A|/|B|)\rho_B$, these elements are compatible under restriction. The same is true of the degree p^n pieces, and so by Quillen's V-isomorphism theorem for complex oriented theories, some power lifts to $\gamma_{ku} \in ku^*(BG)$. Since ku-theory Chern classes reduce to mod p cohomology Chern classes, Duflot's result about associated primes shows that γ_{ku} is regular on mod p cohomology. $\square$

REMARK 3.1.6. (i) We certainly expect this result holds more generally, and note that any extra elements of $H^0_I(ku^*(BG))$ are I-power torsion but also $(v, |G|)$-torsion.

(ii) We warn that an example of Pakianathan's [49] shows that there need not be a regular element of integral cohomology.

3.2. The universal coefficient theorem.

To understand the duality implied by the local cohomology theorem we begin by discussing the ku universal coefficient theorem as it applies to the space BG.

PROPOSITION 3.2.1. *The universal coefficient theorem for BG takes the form of a short exact sequence*

$$0 \longrightarrow \text{Ext}^2_{ku_*}(\Sigma^2 S, ku_*) \longrightarrow \widetilde{ku}^*(BG) \longrightarrow \text{Ext}^1_{ku_*}(\Sigma \widetilde{P}, ku_*) \longrightarrow 0,$$

3.2. THE UNIVERSAL COEFFICIENT THEOREM.

where S is the $(|G|,v)$-power torsion in $\widetilde{ku}_*(BG)$ and $\widetilde{P} = \widetilde{ku}_*(BG)/S$.

Proof: Let $M = ku_*(BG)$. We may consider the filtration

$$\Gamma_{(|G|,v)}M \subseteq \Gamma_{(|G|)}M \subseteq M.$$

We shall show that each of the three subquotients $M(2) = \Gamma_{(|G|,v)}M = S$, $M(1) = \Gamma_{(|G|)}M/\Gamma_{(|G|,v)}M$ and $M(0) = M/\Gamma_{(|G|)}M$ only gives $\mathrm{Ext}^*_{ku_*}(\cdot, ku_*)$ in a single degree:

$$\mathrm{Ext}^*_{ku_*}(M(i), ku_*) = \mathrm{Ext}^i_{ku_*}(M(i), ku_*).$$

This is clear for $i = 0$, since $\widetilde{ku}_*(BG) = ku_* \oplus \widetilde{ku}_*(BG)$, so that $M(0) = ku_*$, which is free. It also follows that $\Gamma_{(|G|)}M = \widetilde{ku}_*(BG)$, so that $M(1) = \widetilde{P}$. Accordingly, once we know that $M(1)$ and $M(2)$ only contribute Ext in degrees 1 and 2, we will have a short exact sequence

$$0 \longrightarrow \mathrm{Ext}^2_{ku_*}(\Sigma^2 M(2), ku_*) \longrightarrow \widetilde{ku}^*(BG) \longrightarrow \mathrm{Ext}^1_{ku_*}(\Sigma M(1), ku_*) \longrightarrow 0$$

as required.

Since $\widetilde{ku}_*(BG)$ is $|G|$-power torsion it suffices to deal with one prime at a time and we localize at a prime p dividing $|G|$ for the rest of the proof. It is natural to use the injective resolution

$$0 \longrightarrow ku^* \longrightarrow ku^*[1/p, 1/v] \longrightarrow ku^*[1/p]/v^\infty \oplus ku^*[1/v]/p^\infty \longrightarrow ku^*/p^\infty, v^\infty \longrightarrow 0$$

of ku^*. The statement about $M(2)$ then follows directly since it is p and v torsion, and

$$\mathrm{Ext}^2_{ku_*}(\Sigma^2 M(2), ku_*) = \mathrm{Hom}_{ku_*}(\Sigma^2 M(2), ku_*/|G|^\infty, v^\infty).$$

Now consider $M(1)$. Since it is p-power torsion, its Ext groups are the cohomology of the sequence

$$0 \xrightarrow{d^0} \mathrm{Hom}_{ku_*}(M(1), ku^*[1/v]/p^\infty) \xrightarrow{d^1} \mathrm{Hom}_{ku_*}(M(1), ku^*/p^\infty, v^\infty).$$

For the term in cohomological degree 1, we note first that $M(1)[1/v] = \widetilde{K}_*(BG)$. This is $H^1_{\mathcal{J}}(R(G))$ in each odd degree, and zero in each even degree. Hence we calculate

$$\begin{aligned}\mathrm{Hom}_{ku_*}(M(1), ku^*[1/v]/p^\infty) &= \mathrm{Hom}_{ku_*}(M(1)[1/v], ku^*[1/v]/p^\infty) \\ &= \mathrm{Hom}_{\mathbb{Z}}(M[1/v]_1, \mathbb{Z}/p^\infty)[v, v^{-1}].\end{aligned}$$

This is $(R(V)/(\rho))^\wedge_J$ in each odd degree, and zero in each even degree.

For the term in cohomological degree 2, note first that $M(1)$ is in odd degrees, since it has no v-torsion and $M(1)[1/v]$ is in odd degrees. Evidently $M(1)$ is also bounded below, whilst $ku^*/p^\infty, v^\infty$ is bounded above and hence any map

$$M(1) \longrightarrow ku^*/p^\infty, v^\infty$$

involves only finitely many terms

$$M(1)_{2k+1} \longrightarrow \mathbb{Z}/p^\infty.$$

Finally, since $M(1)$ has no v-torsion the restriction map

$$\begin{aligned}\mathrm{Hom}^{2k+1}_{ku_*}(M(1), ku^*/p^\infty, v^\infty) &\longrightarrow \mathrm{Hom}_{\mathbb{Z}}(M(1)_{2k+1}, (ku^*/p^\infty, v^\infty)_0) \\ &= \mathrm{Hom}_{\mathbb{Z}}(M(1)_{2k+1}, \mathbb{Z}/p^\infty)\end{aligned}$$

is an isomorphism. Since $M(1)_{2k+1}$ maps monomorphically to $M(1)[1/v]_{2k+1}$ it follows that the differential d^1 is surjective. Thus $\mathrm{Ext}^2_{ku_*}(M(1), ku_*) = 0$. and

$$\mathrm{Ext}^{1,2k+1}_{ku_*}(M(1), ku_*) = \mathrm{Hom}_{\mathbb{Z}}([(R(G)/\rho)/p^\infty]/M(1)_{2k+1}, \mathbb{Z}/p^\infty).$$

□

3.3. Local cohomology and duality.

We now want to combine the Universal coefficient theorem of Section 3.2 with the local cohomology theorem of Section 3.1 to obtain a duality statement. This special case illustrates the general phenomena discussed in [**26**] and [**15**]. For this we use the technology of highly structured rings. Thus we write ku for a non-equivariant commutative S^0-algebra representing ku-cohomology. For an equivariant version we use the Elmendorf-May highly structured inflation $\inf_1^G$ of [**17**] to construct the equivariant S^0-algebra $\widehat{ku} = F(EG_+, \inf_1^G ku)$.

We may summarize the discussion in Section 3.2 in the heuristic statement

$$ku^*(BG) = \mathrm{RD}_{ku_*}(ku_*(BG))$$

where $D_{ku_*}(\cdot) = \mathrm{Hom}_{ku_*}(\cdot, ku_*)$ and R denotes the total right derived functor in some derived category. The precise version of this is the S^0-algebra universal coefficient theorem

$$F(BG_+, ku) = F_{ku}(ku \wedge BG_+, ku),$$

which is obtained from the equivariant statement

$$F(EG_+, ku) = F_{ku}(\widehat{ku} \wedge EG_+, ku),$$

by taking G-fixed points, where ku is still the *non-equivariant* S^0-algebra. The expression makes sense since $\widehat{ku}$ is a ku-module.

On the other hand, by the local cohomology theorem, $ku_*(BG)$ can be calculated using a spectral sequence from $H_I^*(ku^*(BG))$, giving the heuristic statement

$$ku_*(BG) = R\Gamma_I(ku^*(BG)).$$

Indeed, in the context of S^0-algebras we may form the homotopy I-power torsion functor Γ_I on the category of strict $\widehat{ku}$-modules. If $I = (x_1, x_2, \ldots, x_r)$ we define $\Gamma_{(x)}\widehat{ku} = \mathrm{fibre}(\widehat{ku} \longrightarrow \widehat{ku}[1/x])$ and

$$\Gamma_I M = \Gamma_{x_1}\widehat{ku} \wedge_{\widehat{ku}} \Gamma_{x_2}\widehat{ku} \wedge_{\widehat{ku}} \cdots \wedge_{\widehat{ku}} \Gamma_{x_r}\widehat{ku} \wedge_{\widehat{ku}} M.$$

The precise version of the heuristic statment is then the equivalence

$$\widehat{ku} \wedge EG_+ \simeq \Gamma_I F(EG_+, \widehat{ku})$$

of G-spectra [**22**, Appendix] and [**23**].

We now combine the two results relating homology and cohomology of BG. We obtain the heuristic statement

$$ku^*(BG) = \mathrm{RD}_{ku_*}(R\Gamma_I(ku^*(BG))).$$

The precise statement is the equivalence

$$F(EG_+, \widehat{ku}) \simeq F_{ku}(\Gamma_I F(EG_+, \widehat{ku}), ku).$$

An algebraic statement is obtained by taking equivariant homotopy, and makes the heuristic statement precise by using appropriate spectral sequences to calculate

3.4. The ku-homology of cyclic and quaternion groups.

the left hand side. This states that the S^0-algebra $F(EG_+, ku)$ is "homotopically Gorenstein". The terminology arises since if a local ring (R, I) of dimension n is Cohen-Macaulay then $H_I^*(R) = H_I^n(R)$, and R is then Gorenstein if and only if $DH_I^n(R)$ is a suspension of R. However, since $ku^*(BG)$ has depth 0, the actual ring theoretic implications are more complicated. They are nonetheless extremely striking in examples, as is described in Sections 3.1, 3.4, 3.5, and, especially remarkable, that of elementary abelian groups given in Section 4.12.

3.4. The ku-homology of cyclic and quaternion groups.

If G is a cyclic group or a generalized quaternion 2-group, the calculation is particularly easy.

LEMMA 3.4.1. *If G is cyclic or generalized quaternion then there is an element y in $ku^2(BG)$ or $ku^4(BG)$ respectively, such that*
$$H_I^0(R) = \mathbb{Z}[v]$$
$$H_I^1(R) = R/(y^\infty) := \operatorname{cok}(R \longrightarrow R[1/y])$$
If $s > 1$ then $H_I^s(R) = 0$.

Proof: For $H_I^0(R)$ we use 3.1.5. For the rest we observe G acts freely on the unit sphere of a suitable representation V and take $y = e_{ku}(V)$. □

REMARK 3.4.2. When G is cyclic, we may view it as embedded in complex numbers and take V to be the natural representation of G on $\mathbb{C}$, so y has its usual meaning. When G is a quaternion group, we may view it as embedded in the quaternions and take V to be the natural representation of G on $\mathbb{H}$, giving $y = q$ in the notation of Section 2.4. Here we may see explicitly that $I = (q, a, b)$ is the radical of (q) since $a^2 = vaq$ and $b^2 = vbq + q_{2^n - 1} - q$ in the case of $Q_{2^{n+2}}$.

PROPOSITION 3.4.3. *If G is cyclic or generalized quaternion then*
$$ku_*(BG) = \mathbb{Z}[v] \oplus \Sigma^{-1} R/(y^\infty).$$

Explicitly,

(1) *If $\widetilde{R} = R(C_n)/(\rho) = \mathbb{Z}[\alpha]/(1 + \alpha + \cdots + \alpha^{n-1})$ then*
$$ku_{2i-1} BC_n = \widetilde{R}/(1 - \alpha)^i$$

(2) $\widetilde{ku_i} BQ_{2^{n+2}} = A_i \oplus B_i$, *where $A_i = B_i = 0$ if i is even, $A_{4k-3} = A_{4k-1} = \mathbb{Z}/2^k \oplus \mathbb{Z}/2^k$, and $B_{4k-1} = B_{4k+1} = \operatorname{Cok}(P^k)$, where P is the companion matrix of the polynomial*
$$-2^{n+2} - f_1 v^2 q - \cdots - f_{2^n - 1} (v^2 q)^{2^n - 1}$$
derived from the regular representation as in Lemma 2.4.3.

The duality is especially simple for these groups because of the absence of v-torsion.

COROLLARY 3.4.4. *If G is cyclic or generalized quaternion, and $\widetilde{R} = \widetilde{ku}^*(BG)$ then*
$$\operatorname{Ext}^i_{ku_*}(\Sigma^{-1} R/(y^\infty), ku_*) = \begin{cases} \widetilde{R} & \text{if } i = 1 \\ 0 & \text{otherwise} \end{cases}$$

Proof of Proposition 3.4.3: The first sentence is obvious from the Gysin sequence, since both groups act freely on the unit sphere $S(V)$ for some representation V and therefore $EG = S(\infty V)$. Only the explicit statement needs proof.

First assume that $G = C_n$. Then $R = ku^*(BG) = ku^*[[y]]/([n](y))$ and the n-series $[n](y) = y\rho$ so the image of R in $R[1/y]$ is $R/(\rho)$. In degree 0 this is the ring $\widetilde{R}$. In positive degree $2i$ (negative codegree $-2i$) the image is $v^k \widetilde{R}$. In contrast, $R[1/y]$ in positive degree $2i$ is $(1/y^i)\widetilde{R}$. Since $v^i = (1-\alpha)^i/y^i$, we see that

$$ku_{2i-1} = H_I^{1,-2i}(R) = \widetilde{R}/(1-\alpha)^i.$$

When $G = Q_{2^{n+2}}$, recall from Lemma 2.4.10 that $R/(\rho)$ in degree $2k$ has an additive basis which is v^k times $\{va, vb, 1, v^2q, (v^2q)^2, \ldots, (v^2q)^{2^n-1}\}$. The same lemma shows that $R[1/q]_{4k} = R^4/q^{k+1}$ has basis $(1/q^{k+1})$ times $\{vaq, vbq, q, v^2q^2, \ldots, q(v^2q)^{2^n-1}\}$ and $R[1/q]_{4k+2} = R^2/q^{k+1}$ has basis $(1/q^{k+1})$ times $\{a, b, vq, v^3q^2, \ldots, vq(v^2q)^{2^n-1}\}$. If we filter $R/(\rho)$ by putting the summands involving a and b into filtration 0 and the rest into filtration 1, then both multiplication by v and q are filtration preserving, and we can compute $H_I^1(R)$ as the direct sum of two cokernels: one involving a and b, and the other involving only powers of q. For the first part note that $v^2 a = 2a/q$ and $v^2 b = 2b/q$ modulo terms involving only powers of q. Hence in degree $4k-2$ the map $R/(\rho) \longrightarrow R[1/q]$ is

$$<v^{2k}a, v^{2k}b> \;=\; <2^k a/q^k, 2^k b/q^k> \;\longrightarrow\; <a/q^k, b/q^k>$$

with cokernel $\mathbb{Z}/2^k \oplus \mathbb{Z}/2^k$. In degree $4k$ we have v times this with the same cokernel. This accounts for the summands A_i.

For the remainder, recall from Lemma 2.4.3 that the regular representation is a polynomial of degree 2^n in the Euler class $v^2 q$ of a faithful representation:

$$\rho = 2^{n+2} + f_1(v^2 q) + \cdots + f_{2^n-1}(v^2q)^{2^n-1} + (v^2q)^{2^n}.$$

Since ρ is zero in $R[1/q]$, there we have

$$(v^2q)^{2^n} = -2^{n+2} - f_1(v^2q) - \cdots - f_{2^n-1}(v^2q)^{2^n-1}.$$

In degree $4k$ the inclusion of $R/(\rho)$ into $R[1/q]$ is the inclusion of

$$v^{2k} <1, v^2 q, \ldots, (v^2q)^{2^n-1}> \;\longrightarrow\; (1/q^k) <1, v^2 q, \ldots, (v^2q)^{2^n-1}>$$

and in degree $4k+2$ we have v times this. It is not hard to see that this is exactly the group presented by the k-th power of the companion matrix of the polynomial expressing the highest power of v^2q in terms of lower ones as above. $\square$

REMARK 3.4.5. For Q_8 the relation obtained by setting the regular representation to zero and inverting q is $v^4 q^2 = -8 + 6v^2 q$, so that we have

$$B_{4k-1} = B_{4k+1} = \mathbb{Z}_{2^{2k+1}} \oplus \mathbb{Z}_{2^{k-1}},$$

the group presented by the k-th power of the companion matrix

$$\begin{pmatrix} 0 & -8 \\ 1 & 6 \end{pmatrix}.$$

REMARK 3.4.6. For a cyclic group of prime order p, the result is especially simple. In this case the ring $\widetilde{R} = \mathbb{Z}[\alpha]/(1 + \alpha + \cdots + \alpha^{p-1})$ is the ring generated by a primitive p^{th} root of unity in the complex numbers. Here we can calculate that

$(1-\alpha)^{p-1}$ is p times a unit. This, together with the calculation of $\widetilde{R}/(1-\alpha)^i$ for $i=1,\ldots,p-2$, gives

$$\widetilde{ku}_n BC_p = \begin{cases} (\mathbb{Z}/p^{j+1})^s \oplus (\mathbb{Z}/p^j)^{p-1-s} & \text{if } n=2j(p-1)+2s-1 \\ & \text{with } 0<s\leq p-1 \\ 0 & \text{otherwise} \end{cases}$$

For a general cyclic group, there will be sets of summands with different rates of growth corresponding to the factorization of $1+\alpha+\cdots+\alpha^{n-1}$ into cyclotomic polynomials. For example,

$$\widetilde{ku}_n BC_{p^2} = A_n \oplus B_n$$

where

$$A_n = \begin{cases} (\mathbb{Z}/p^{j+2})^s \oplus (\mathbb{Z}/p^{j+1})^{p-1-s} & \text{if } n=2j(p-1)+2s-1 \\ & \text{with } 0<s\leq p-1 \\ 0 & \text{otherwise} \end{cases}$$

and

$$B_n = \begin{cases} (\mathbb{Z}/p^{j+1})^s \oplus (\mathbb{Z}/p^j)^{p^2-p-1-s} & \text{if } n=2j(p^2-p)+2s+2p-3 \\ & \text{with } 0<s\leq p^2-p \\ 0 & \text{otherwise.} \end{cases}$$

3.5. The ku-homology of BD_8.

In this section we shall compute ku_*BD_8 as a module over

$$\begin{aligned} R &= ku^*BD_8 \\ &= ku^*[a,b,d]/(v^4d^3 - 6v^2d^2 + 8d, va^2 - 2a, vb^2 - 2b, ab - (b^2 - vbd), \\ &\quad 2ad, vad, 2bd - v^2bd^2, vbd - (4d - v^2d^2)) \end{aligned}$$

where $a,b \in R^2$ and $d \in R^4$ (Theorem 2.5.5). We shall use the short exact sequence of R-modules

$$0 \longrightarrow T \longrightarrow R \longrightarrow Q \longrightarrow 0,$$

where Q is the image of R in K^*BD_8 and $T = \Gamma_{(2,v)}R = \Gamma_{(2)}R = \Gamma_{(v)}R$. The R action on T factors through $P = R/(2,v)R = \mathbb{F}_2[a,b,d]/(ab+b^2)$, and it is easy to see that T is the free P-module generated by $ad \in R^6$. We write $M^\vee = \operatorname{Hom}_{\mathbb{F}_2}(M, \mathbb{F}_2)$ for the $\mathbb{F}_2$-dual of any P-module: note that if M is in positive degrees, $M^\vee$ is in negative degrees, and vice versa, and that $M^\vee$ is again a P-module.

The remainder of this section will be devoted to the proof of the following theorem.

THEOREM 3.5.1. *As an R-module, $\widetilde{ku}_*BD_8 = \widetilde{ku}_{odd}BD_8 \oplus \widetilde{ku}_{even}BD_8$.*
  (1) $\widetilde{ku}_{even}BD_8 = \Sigma^{-2}H_I^2(R) = \Sigma^2 P^\vee$. *Additively*, $\widetilde{ku}_{2i}BD_8 = (\mathbb{Z}/2)^i$.
  (2) $\widetilde{ku}_{odd}BD_8 = \Sigma^{-1}H_I^1(R)$, *with additive generators a_i, b_i, c_i, and d_i in $ku_{2i-1}BD_8$ for $i > 0$.*
    (a) $ku_1BD_8 = (\mathbb{Z}/2)^2 = \langle a_1\rangle \oplus \langle b_1\rangle$, *with $d_1 = a_1$ and $c_1 = 0$.*
    (b) $ku_3BD_8 = (\mathbb{Z}/4)^3 = \langle a_2\rangle \oplus \langle b_2\rangle \oplus \langle d_2\rangle$ *and $c_2 = 2a_2 + 2d_2$.*
    (c) $ku_5BD_8 = (\mathbb{Z}/8)^3 = \langle a_3\rangle \oplus \langle b_3\rangle \oplus \langle d_3\rangle$ *and $c_3 = 4a_3 + 4d_3$.*

	v	a	b	d
a_i	$2a_{i+1}$	a_{i-1}	$b_{i-1} - 2c_{i-1}$	0
b_i	$2(b_{i+1} - \epsilon_{i+1}c_{i+1})$	$b_{i-1} - 2c_{i-1}$	$b_{i-1} + 2\epsilon_i c_{i-1}$	$2c_{i-2}$
c_i	$2c_{i+1}$	0	$(1+\epsilon_i)c_{i-1}$	c_{i-2}
d_i	$2d_{i+1} - \epsilon_i c_{i+1}$	0	c_{i-1}	d_{i-2}

FIGURE 3.1. The R-module structure of $ku_{\mathrm{odd}} BD_8$. We let $\epsilon_{2i} = 0$ and $\epsilon_{2i+1} = 1$.

(d) For $i \geq 4$, $ku_{2i-1}BD_8 = (\mathbb{Z}/2^i)^3 \oplus A_{2i-1} = <a_i> \oplus <b_i> \oplus <d_i> \oplus A_{2i-1}$, where
   (i) $A_{4n-1} = (\mathbb{Z}/2^{n-1}) = <c_{2n} + 2^n(a_{2n} + d_{2n})>$
   (ii) $A_{4n+1} = (\mathbb{Z}/2^{n-1}) = <c_{2n+1} + 2^{n+1}(a_{2n+1} + d_{2n+1})>$

The R-module structure is given in Figure 3.1.

There is a simple heuristic which neatly describes the summands A_{2i-1}. First note that $2^n c_{2n} = 0$ and $2^n c_{2n+1} = 0$. For any element x, write $s(x)$ for $2^{i-1}x$ if x has order 2^i. Then $s(c_i) = s(a_i) + s(d_i)$. Thus, by adding the appropriate multiple of $a_i + d_i$ to c_i we obtain a summand whose order is half the order of c_i.

Proof: Since R is concentrated in even degrees, ku_*BD_8 splits as the sum of the even and odd degree parts. We shall first calculate the local cohomology of T and of Q, then use the long exact sequence of local cohomology to calculate the local cohomology of R. We will show that $H_I^0(R) = ku^* \cdot \rho$, which is a direct summand of ku_*BD_8. Since R has dimension 2, it follows that the local cohomology spectral sequence must collapse to give isomorphisms $\widetilde{ku_{\mathrm{even}}}BD_8 = H_I^2(R)$ and $ku_{\mathrm{odd}}BD_8 = H_I^1(R)$.

We start with T, which we have already observed is isomorphic to $\Sigma^6 P$, where $P = \mathbb{F}_2[a,b,d]/(ab+b^2)$. (Our suspensions are cohomological; that is, $T^{n+6} = (\Sigma^6 P)^{n+6} = P^n$.) It is easy to calculate directly that $H_I^0(T) = H_I^1(T) = 0$ and $H_I^2(T) = \Sigma^6 H_I^2(P) = \Sigma^6 \Sigma^{-4} P^\vee = \Sigma^2 P^\vee$. Note that P is isomorphic to $H^*(BD_8; \mathbb{F}_2)$ under a degree doubling isomorphism. We could give an alternative calculation of the local cohomology by arguing that $H^*(BD_8)$ is Cohen-Macaulay and hence, from the local cohomology theorem, Gorenstein.

The local cohomological dimension of Q, like that of $K^*BD_8 = R[1/v]$, is 1 (see Lemma 3.5.5 for an explicit proof). Therefore, the long exact sequence for H_I^* gives an isomorphism $H_I^0(R) = H_I^0(Q)$, which we shall show is $ku^* \cdot \rho$, just as in 3.1.5, and an exact sequence

$$0 \longrightarrow H_I^1(R) \longrightarrow H_I^1(Q) \longrightarrow H_I^2(T) \longrightarrow H_I^2(R) \longrightarrow 0$$

Observe that the map in the middle will be completely determined by its behaviour in the bottom degree (top codegree) because the dual of $H_I^2(T)$ is monogenic.

We next compute $H_I^*(Q)$. It will help to have an explicit description of Q.

LEMMA 3.5.2. $Q = ku^*[a,b,d]/(ad, va^2 - 2a, vb^2 - 2b, b(a-b+vd), d(4-vb-v^2d))$, which is a free $\mathbb{Z}_2^\wedge$-module on the following basis:

codegree	basis
$4n+2$	$a^{2n+1}, vd^{n+1}, bd^n, b^{2n+1}$
$4n$	$a^{2n}, d^n, vbd^n, b^{2n}$
6	a^3, vd^2, bd, b^3
4	a^2, d, vbd, b^2
2	a, vd, v^2bd, b
0	$1, va, v^2d, v^3bd, vb$
$-2i, i>0$	v^i times the basis for Q^0

REMARK 3.5.3. Since Q is a subring of K^*BD_8, computations in Q can be done in the representation ring. We will write elements of the representation ring by giving their characters, where the character table has columns identity, center, rotation, reflection, reflection, respectively. The images in the character ring of the generators are:

$$va = (00022)$$
$$vb = (00202)$$
$$v^2d = (04200)$$

DEFINITION 3.5.4. Let $y = d + a^2 \in Q^4$.

LEMMA 3.5.5. The radical of (y) is I, and hence $H_I^* = H_{(y)}^*$. There is an exact sequence

$$0 \longrightarrow H_I^0(Q) \longrightarrow Q \longrightarrow Q[1/y] \longrightarrow H_I^1(Q) \longrightarrow 0.$$

Now, $v^2y = (04244)$ in the character ring, and is therefore nonzero on any element which is not a multiple of the regular representation $\rho = (80000)$. This gives us a practical method for completing the computation.

COROLLARY 3.5.6. $H_I^0(Q) = ku^* \cdot \rho$, the free ku^*-module generated by the regular representation $\rho = 8 - 4va - 2v^2d - v^3bd$. There is a short exact sequence

$$0 \longrightarrow Q/(\rho) \longrightarrow Q[1/y] \longrightarrow H_I^1(Q) \longrightarrow 0.$$

This accounts for $H_I^0(R) = H_I^0(Q)$. To describe $H_I^1(Q)$, let us temporarily use the abbreviation $H^n := H_I^{1,n}(Q)$. Since multiplication by y is an isomorphism in $Q[1/y]$, and $Q^i = Q[1/y]^i$ for $i \geq 4$, we find that we can compute

$$H^{-4n} = Q^4/(y^{n+1}Q^{-4n})$$
$$H^{-4n+2} = Q^6/(y^{n+1}Q^{-4n+2}).$$

Then H^{-4n} is spanned by

$$\widetilde{a}_{2n} = \frac{a^2}{y^{n+1}} \qquad \widetilde{b}_{2n} = \frac{b^2}{y^{n+1}} \qquad \widetilde{c}_{2n} = \frac{vbd}{y^{n+1}} \qquad \widetilde{d}_{2n} = \frac{d}{y^{n+1}}$$

and H^{-4n+2} by

$$\widetilde{a}_{2n-1} = \frac{a^3}{y^{n+1}} \qquad \widetilde{b}_{2n-1} = \frac{b^3}{y^{n+1}} \qquad \widetilde{c}_{2n-1} = \frac{bd}{y^{n+1}} \qquad \widetilde{d}_{2n-1} = \frac{vd^2}{y^{n+1}}.$$

PROPOSITION 3.5.7. The top local cohomology $H^* = H_I^{1,*}(Q)$ is

- $H^2 = \mathbb{Z}/2 = <\widetilde{b}_{-1}>$, $\widetilde{c}_{-1} = \widetilde{b}_{-1}$, and $\widetilde{a}_{-1} = \widetilde{d}_{-1} = 0$.
- $H^0 = \mathbb{Z}/2 \oplus \mathbb{Z}/2 = <\widetilde{a}_0, \widetilde{b}_0>$, $\widetilde{c}_0 = 0$, and $\widetilde{d}_0 = \widetilde{a}_0$.

- $H^{-2} = \mathbb{Z}/4 \oplus \mathbb{Z}/4 \oplus \mathbb{Z}/2 = <\tilde{a}_1, \tilde{b}_1, \tilde{c}_1>$ and $\tilde{d}_1 = 2\tilde{a}_1$.
- $H^{-4} = \mathbb{Z}/8 \oplus \mathbb{Z}/8 \oplus \mathbb{Z}/8 = <\tilde{a}_2, \tilde{b}_2, \tilde{d}_2>$ and $\tilde{c}_2 = 4(\tilde{a}_2 + \tilde{d}_2)$.

and in codegrees less than -4,

- $H^{-4n} = (\mathbb{Z}/2^{2n+1})^3 \oplus \mathbb{Z}/2^{n-1}$ generated by $\tilde{a}_{2n}$, $\tilde{b}_{2n}$, $\tilde{d}_{2n}$, and $\tilde{c}_{2n} + 2^{n+1}(\tilde{a}_{2n} + \tilde{d}_{2n})$, respectively.
- $H^{-4n-2} = (\mathbb{Z}/2^{2n+2})^2 \oplus \mathbb{Z}/2^{2n+1} \oplus \mathbb{Z}/2^n$ generated by $\tilde{a}_{2n+1}$, $\tilde{b}_{2n+1}$, $\tilde{d}_{2n+1}$, and $\tilde{c}_{2n+1} + 2^{n+1}\tilde{a}_{2n+1} + 2^n\tilde{d}_{2n+1}$, respectively.

The same heuristic we used to describe the generators of the small summands of ku_*BD_8 works here as well. We have $s(\tilde{c}_i) = s(\tilde{a}_i) + s(\tilde{d}_i)$. Thus, by adding the appropriate (not equal) multiples of $\tilde{a}_i$ and $\tilde{d}_i$ to $\tilde{c}_i$ we obtain a summand whose order is half the order of $\tilde{c}_i$.

Next we wish to determine the map $H^1_I(Q) \longrightarrow H^2_I(T)$. We shall write the elements of $H^2_I(T) = P^\vee$ dual to $a^i d^j$ and $b^i d^j$, respectively, as $1/a^i d^j$ and $1/b^i d^j$. Then we have

PROPOSITION 3.5.8. *The homomorphism $H^1_I(Q) \longrightarrow H^2_I(T)$ is*

$$\tilde{a}_i \mapsto 1/b^{i+1}$$
$$\tilde{b}_i \mapsto 1/a^{i+1} + 1/b^{i+1}$$
$$\tilde{c}_{2i} \mapsto 0 \quad \text{and} \quad \tilde{c}_{2i-1} \mapsto 1/d^i$$
$$\tilde{d}_{2i} \mapsto 1/bd^i \quad \text{and} \quad \tilde{d}_{2i-1} \mapsto 0$$

Proof: The bottom class must map nontrivially by connectivity and collapse of the local cohomology spectral sequence. Then the R-module structure of $H^1_I(Q)$ and $H^2_I(T)$ forces the rest of the map. □

The cokernel of this map is easily computed.

PROPOSITION 3.5.9. *The map $H^2_I(T) \longrightarrow H^2_I(R)$ is dual to the inclusion of the ideal $(ad) \subset P = \mathbb{F}_2[a,b,d]/(ab+b^2)$ and this ideal is isomorphic to P, so $H^2_I(R) = P^\vee$, with bottom class in degree 4.*

The kernel is also easy to compute, though it requires more work to describe. Let us define

$$a_i = 2\tilde{a}_i$$
$$b_i = 2\tilde{b}_i$$
$$c_{2i} = \tilde{c}_{2i} \quad \text{and} \quad c_{2i-1} = 2\tilde{c}_{2i-1}$$
$$d_{2i} = 2\tilde{d}_{2i} \quad \text{and} \quad d_{2i-1} = \tilde{d}_{2i-1}$$

Then clearly $H^{1,2i}_I(R)$ is generated by a_i, b_i, c_i and d_i, and we have completed the calculation of $H^*_I(R)$.

PROPOSITION 3.5.10. $H_I^1(R)$ is as follows:

$$H_I^{1,-4n}(R) = (\mathbb{Z}/2^{2n})^3 \oplus \mathbb{Z}/2^{n-1}$$
$$= \langle a_{2n} \rangle \oplus \langle b_{2n} \rangle \oplus \langle d_{2n} \rangle \oplus \langle c_{2n} + 2^n(a_{2n} + d_{2n}) \rangle$$
$$H_I^{1,-4n}(R) = (\mathbb{Z}/2^{2n+1})^3 \oplus \mathbb{Z}/2^{n-1}$$
$$= \langle a_{2n+1} \rangle \oplus \langle b_{2n+1} \rangle \oplus \langle d_{2n+1} \rangle$$
$$\oplus \langle c_{2n+1} + 2^{n+1}(a_{2n+1} + d_{2n+1}) \rangle$$

The only thing left to compute in order to finish the proof of Theorem 3.5.1 is the action of R, and this follows easily from the R action in $Q[1/y]$. □

The duality obtained by combining the local cohomology spectral sequence and the universal coefficient spectral sequence says roughly that the v-torsion submodules of $ku^*(BD_8)$ and $ku_*(BD_8)$ are dual and that their v-torsion-free quotients are dual. To be precise, first note that each contains a copy of ku^*, and that these are dual by

$$ku_* = H_I^0(ku^*BD_8) = ku^* \cdot \rho$$

and

$$ku^* = \mathrm{Hom}_{ku_*}(ku_*BD_8, ku_*).$$

The interest lies in the reduced groups. We have the short exact sequence of ku^*BD_8 modules

$$(4) \qquad 0 \longrightarrow T \longrightarrow \widetilde{ku}^*BD_8 \longrightarrow \widetilde{Q} \longrightarrow 0.$$

The local cohomology spectral sequence collapses to a short exact sequence

$$0 \longrightarrow \Sigma^{-2}T^\vee \longrightarrow \widetilde{ku}_*BD_8 \longrightarrow H_I^1(\widetilde{ku}^*BD_8) \longrightarrow 0$$

and this is exactly the short exact sequence expressing $\widetilde{ku}_*BD_8$ as the extension of the v-torsion submodule by the v-torsion-free quotient. Note that the top of T is in degree -6, so that the bottom of $\Sigma^{-2}T^\vee$ is in degree 4. Theorem 3.2.1 tells us that the universal coefficient spectral sequence degenerates to the short exact sequence

$$0 \longrightarrow \mathrm{Ext}^2_{ku_*}(T^\vee, ku_*) \longrightarrow \widetilde{ku}^*(BD_8) \longrightarrow \mathrm{Ext}^1_{ku_*}(\Sigma H_I^1(\widetilde{ku}^*BD_8), ku_*) \longrightarrow 0.$$

One may check that the Ext^1 group is v-torsion free, and therefore this coincides with our original extension (4), leading to the duality statements which follow.

PROPOSITION 3.5.11. The short exact sequence for $\widetilde{ku}_*(BD_8)$ from the local cohomology spectral sequence is dual to the short exact sequence for $\widetilde{ku}^*(BD_8)$ from the universal coefficient theorem. More precisely,

$$\mathrm{Ext}^i_{ku_*}(T^\vee, ku_*) = \begin{cases} T & \text{if } i = 2 \\ 0 & \text{otherwise} \end{cases}$$

and

$$\mathrm{Ext}^i_{ku_*}(\Sigma H_I^1(\widetilde{ku}^*BD_8), ku_*) = \begin{cases} \widetilde{Q} & \text{if } i = 1 \\ 0 & \text{otherwise.} \end{cases}$$

Even more striking are the corresponding dualities in Section 4.12 for elementary abelian groups.

3.6. Tate cohomology.

As with any other theory, there is a Tate cohomology associated to ku [30] and we have the norm sequence

$$\cdots \longrightarrow ku_*(BG) \xrightarrow{N} ku^*(BG) \longrightarrow t(ku)_G^* \longrightarrow \Sigma ku_*(BG) \longrightarrow \cdots.$$

It is of interest to calculate the Tate cohomology, partly for its own sake, and partly because of the information it encodes about the relationship between homology and cohomology. For ku, the Tate cohomology contains most of the information about both homology and cohomology in the sense that the norm map N is nearly trivial.

LEMMA 3.6.1. *The image of the norm map $ku_*(BG) \longrightarrow ku^*(BG)$ is the free ku^*-module on the regular representation in $R(G)_J^\wedge = ku^0(BG)$.*

Proof: First note that $ku_*(BG)$ is in positive degrees, so in the range where the image of the norm is non-trivial, $ku^*(BG) \longrightarrow K^*(BG)$ is an isomorphism. The result therefore follows from the corresponding fact for periodic K-theory. □

This means that we have an extension

$$0 \longrightarrow ku^*(BG)/(\rho) \longrightarrow t(ku)_G^* \longrightarrow \Sigma \widetilde{ku}_*(BG) \longrightarrow 0.$$

This can conveniently be displayed by superimposing the Adams spectral sequence for $ku^*(BG)$ (which occupies the part of the upper half-plane specified by $s \geq (t-s)/2$) and the Adams spectral sequence for $\Sigma \widetilde{ku}_*(BG)$ (which occupies the part of the upper half-plane specified by $s < (t-s)/2$). The additive extensions can be added to this.

For periodic K-theory the extension is as non-trivial as possible in the sense that $t(K)_G^*$ is a rational vector space, terms of which arise out of $0 \longrightarrow \mathbb{Z}_p^\wedge \longrightarrow \mathbb{Q}_p \longrightarrow \mathbb{Z}/p^\infty \longrightarrow 0$. This is characteristic of one dimensional rings as shown in [24], and the argument of Benson and Carlson [7] shows that for rings of depth 2 or more, the extension is multiplicatively nearly trivial. However $ku^*(BG)$ is of depth zero, and it appears that the Tate extension is highly non-trivial.

For groups which act freely on a sphere, we can solve this extension problem by James periodicity [30, 16.1] :

(5) $$t(ku)^G = \varprojlim(ku \wedge \Sigma P_{-n}^\infty)$$

where P_{-n}^∞ is the associated stunted projective space.

THEOREM 3.6.2. (1) *If $G = C_n$, the cyclic group of order n, then*

$$t(ku)_*^G = ku^*[[y]][1/y]/(\rho) = ku^*[[y]][1/y]/([n](y)).$$

(2) *If $Q = Q_{2^{n+2}}$, the quaternion group of order 2^{n+2}, then*

$$t(ku)_*^Q = ku^*(BQ)[1/q]$$

where $q \in ku^4(BQ)$ is the Euler class of a faithful 2-dimensional representation.

3.6. TATE COHOMOLOGY.

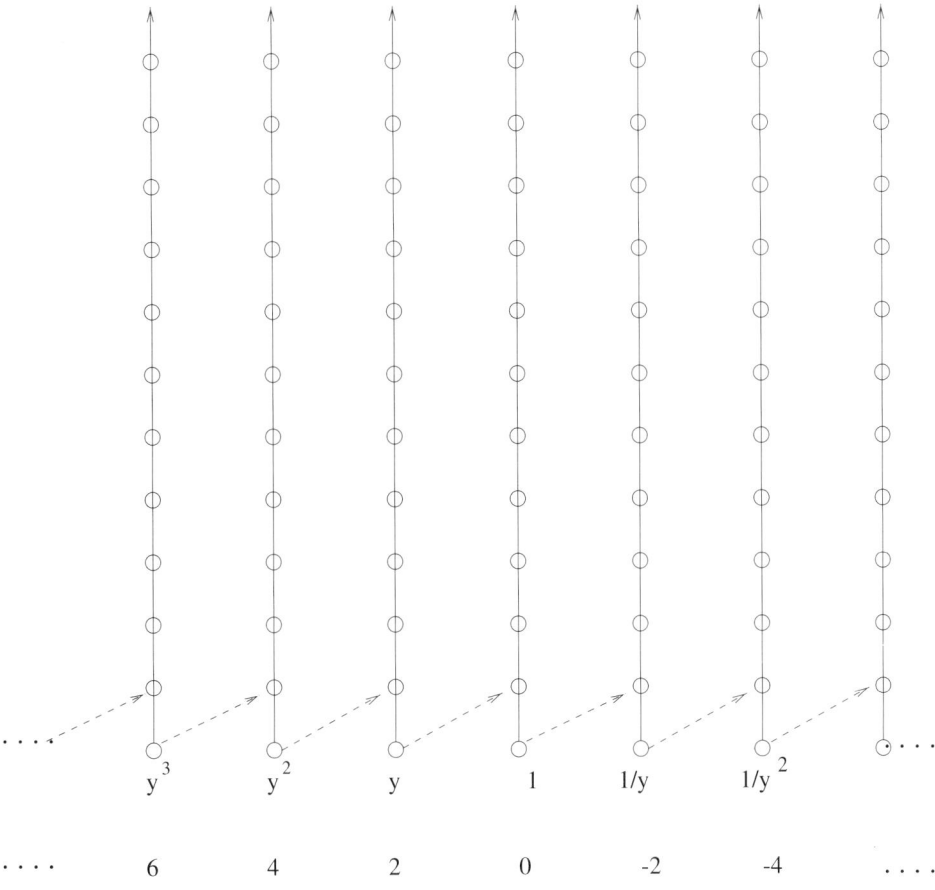

FIGURE 3.2. $t(ku)^*_{C_2} = \mathbb{Z}_2^\wedge[y, 1/y]$ with $v = 2/y$

Proof: James periodicity tells us that the limit (5) is simply $ku^*BC[1/y]$ if C is cyclic, and is $ku^*BQ[1/q]$ if Q is generalized quaternion. Here $y \in ku^2BC$ and $q \in ku^4BQ$ are as in Lemma 3.4.1. □

In these cases, the theorem tells us that $t(ku)^G_*$ is simply the periodic extension of ku^2BC with period 2, when C is cyclic, and of ku^2BQ and ku^4BQ with period 4 when Q is generalized quaternion. (See figures 2.5, 2.6, 2.7, and 2.12.)

This has a deceptively simple form when C is cyclic of prime order (see Figure 3.2), as shown by Davis and Mahowald.

THEOREM 3.6.3. ([**12**]) *If p is a prime then*
$$t(ku)^{C_p} = \bigvee_{j=0}^{p-2} \Sigma^{2j} \prod_{i=-\infty}^{\infty} \Sigma^{2i} H\mathbb{Z}_p^\wedge.$$

Proof: Recall that p-locally, $ku \simeq l \vee \Sigma^2 l \vee \cdots \vee \Sigma^{2(p-2)} l$. The Adams spectral sequence shows that $\pi_{2i} t(l)^{C_p} = \mathbb{Z}_p^\wedge$, generated by a class detected in cohomology, as this is true of $l^2 BC_p$. Let $y_i : t(l)^{C_p} \longrightarrow \Sigma^{2i} H\mathbb{Z}_p^\wedge$ be such a cohomology class.

The product over all i of the y_i induces an isomorphism in homotopy and is therefore an equivalence. The result for ku follows. □

This splitting into Eilenberg-MacLane spectra is quite exceptional, however, and is probably confined to cyclic groups of prime order.

PROPOSITION 3.6.4. *If $G = C_{p^k}$ for $k > 1$, Q_{2^n} for $n \geq 3$, or $SL(2,\ell)$, then $t(ku)^G$ is not a generalized Eilenberg-MacLane spectrum.*

Proof: The inverse limit description (5) of $t(ku)^G$ applies when G is C_{p^k} or Q_{2^n}. If G is $SL(2,\ell)$ then transfer and restriction allow us to split the 2-localization of $t(ku)^G$ off $t(ku)^{Q_{2^n}}$, and to identify the relevant part of the inverse sequence for $t(ku)^{Q_{2^n}}$. Thus, in each case we have a description of the G-fixed points of the Tate theory as an inverse limit. We shall show it is not a generalized Eilenberg-MacLane spectrum (GEM) by using Goerss's theorem on the homology of inverse limits. Goerss shows in [20] that the homology of the inverse limit is the inverse limit of the homologies in the category of comodules, (the inverse sequence is Mittag-Löffler), and that this can be calculated by taking the inverse limit in the category of vector spaces and restricting to those elements whose coproducts have only a finite number of terms.

The comodule $H_*(ku \wedge P_{-n})$ is the extended comodule $H_*(ku)\square_{E(1)}H_*(P_{-n})$. For $G = C_{p^k}$, $H_*(P_{-n})$ is a trivial $E(1)$-comodule. It follows that the homology of the inverse limit $t(ku)^G$ is the finite coproduct part of the product of one copy of $H_*(ku)$ for each integer. Suppose $p = 2$. Then $H_*(ku)$ is $\mathbb{F}_2[\xi_1^4, \xi_2^2, \xi_3, \ldots]$. It follows that the homology of the inverse limit has the property that the coproduct of any element of $B = H_* t(ku)^G$ lies in $\mathbb{F}_2[\xi_1^4, \xi_2^2, \xi_3, \ldots] \otimes B$.

The homotopy of $t(ku)^G$, computed in Theorem 3.6.2, is a sum of copies of $\mathbb{Z}_2^\wedge$ in each even dimension. Thus, if $t(ku)^G$ is a GEM, it is a product of $H\mathbb{Z}_2^\wedge$'s. Now $H_* H\mathbb{Z}_2^\wedge = \mathbb{F}_2[\xi_1^2, \xi_2, \xi_3, \ldots]$, so the homology of such a GEM will have elements whose coproduct contains terms of the form $\xi_1^2 \otimes x$. Since $H_*(t(ku)^G)$ does not, $t(ku)^G$ cannot be a GEM.

When $p > 2$, a similar argument applies with ξ_1^2 replaced by τ_1. When $G = SL(2,\ell)$, we also have that the relevant part of P_{-n} has homology which is a trivial $E(1)$-comodule, so the same argument applies. Finally, since $t(ku)^{SL(2,\ell)}_{(2)}$ splits off $t(ku)^{Q_{2^n}}$, the latter cannot be a GEM either. □

REMARK 3.6.5. For real connective K-theory, the issue is more easily resolved, because there are often nontrivial multiplications by η. While $t(ko)^G$ is a GEM for G cyclic of prime order ([12]), Bayen and Bruner ([6]) show that $t(ko)^G$ is not a GEM for $G = Q_8$, and similar calculations show this for any quaternion or dihedral group.

CHAPTER 4

The ku-homology and ku-cohomology of elementary abelian groups.

In this chapter we discuss elementary abelian groups, mostly at the prime 2. These are the only class of examples of higher rank that we consider, but this suffices to illustrate the complexity of the structure.

In Section 4.2 the Adams spectral sequence is used to calculate the cohomology ring $ku^*(BV)$ for an arbitrary elementary abelian group V of rank r. As usual, this is a mixture of the one dimensional part from periodic K-theory and an r-dimensional part from mod p cohomology. However both of these pieces are slightly modified from their simple form and the way they are stuck together is also interesting.

The succeeding sections are the most complicated bits of commutative algebra in this paper. The bulk of the work is involved in calculating the local cohomology of $ku^*(BV)$, but it shows that it is a very remarkable module. Still more striking is the way that although there are many differentials in the local cohomology spectral sequence, they are all forced for formal reasons, and all but three columns are wiped out before the E^∞ term. This allows us to calculate $ku_*(BV)$ up to a single extension and thereby to investigate the duality.

The results in this chapter are more cumulative than in previous cases, so we start with Section 4.1 summarizing the results and the organization of the rest of the chapter.

4.1. Description of results.

The aim of this chapter is to have a detailed understanding of the homology and cohomology of BV for an elementary abelian group V of rank r. The basis for this is the calculation in Section 4.2 of the cohomology ring $R = ku^*(BV)$. First one may understand the mod p cohomology ring of BV over $E(1)$ and, by Ossa's splitting theorem, the Adams spectral sequence for $ku^*(BV)$ collapses. One may construct enough elements to deduce that there is a short exact sequence

$$0 \longrightarrow T \longrightarrow ku^*(BV) \longrightarrow Q \longrightarrow 0$$

of R-modules. Here T is the ideal of (p,v)-torsion elements of R, and Q has no p or v torsion. In fact Q is the image of $R = ku^*(BV)$ in $K^*(BV)$ and one may describe it explicitly. Furthermore the ring homomorphism

$$R \longrightarrow K^*(BV) \times H^*(BV; \mathbb{F}_p)$$

is injective, the image in $K^*(BV)$ is the Rees ring Q, and T maps monomorphically into the second factor with an image that may be described explicitly as an ideal in $H^*(BV; \mathbb{F}_p)$.

Now let $p = 2$ and consider the calculation of the homology of BV from the augmented cohomology ring of BV using the local cohomology theorem. The principal advantage of this approach is that it is essentially independent of a basis of V. Accordingly, the combinatorics and homological algebra is intrinsic to the problem. By contrast, in higher rank examples the combinatorics arising by conventional means involves arbitrary choices and appears intricate and mysterious. Finally, the homological and combinatorial structures emerging here suggest the lines of the calculation for higher chromatic periodicities, such as $BP\langle n \rangle$ for $n \geq 2$. We intend to investigate this further elsewhere. The calculation of $ku_*(BV)$ is a substantial enterprise and occupies Sections 4.3 to 4.11; we describe the argument in outline before explaining the contents in more detail.

We let
$$I = (y_1, y_2, \ldots, y_r)$$
denote the ideal generated by the Euler classes y_i, and note that by [23] this has radical equal to the augmentation ideal $\ker(R = ku^*(BV) \longrightarrow ku^*)$. The local cohomology theorem (in this case an immediate consequence of the fact that we may use $S(\infty\alpha_1) \times S(\infty\alpha_2) \times \cdots \times S(\infty\alpha_r)$ as a model for EV) states that there is a spectral sequence
$$E^2_{p,q} = H_I^{-p,-q}(R) \Rightarrow ku_{p+q}(BV).$$
We view this as a spectral sequence concentrated in the first $r + 1$ columns of the left half-plane. It therefore has $r - 1$ differentials, $d^2, d^3, \ldots, d^r$, with
$$d^i : E^i_{p,q} \longrightarrow E^i_{p-i,q+i-1}.$$
This is a spectral sequence of R-modules. We emphasize that the E^2 term is a functor of R, so exposes intrinsic structure, and the entire spectral sequence is natural in V.

Accordingly, our first task is to calculate the local cohomology groups $H_I^*(R)$. First we remark that the submodule T of $(2, v)$-torsion is equal to the submodule of 2-torsion, or equally, to the submodule of v-torsion (4.2.3). We thus begin with the short exact sequence
$$0 \longrightarrow T \longrightarrow R \longrightarrow Q \longrightarrow 0,$$
where T is the v-torsion and Q is the Rees ring (the image of R in $K^*(BV)$). It turns out that T is a direct sum of $r - 2$ submodules $T_2, T_3, \ldots, T_r$, with T_i of projective dimension $r - i$.

Before proceeding, we outline the shape of the answer. One piece of terminology is useful. *To have property P in codimension i*, means that any localization at a prime of height i has property P. Thus Gorenstein in codimension 0 means the localization at any minimal prime is Gorenstein, Gorenstein in codimension 1 is stronger, and Gorenstein in codimension r is the same as Gorenstein (since Gorenstein is *defined* as a local property).

We shall see that Q has local cohomology only in degrees 0 and 1 (it is of dimension 1) whilst T_i has depth i and only has local cohomology in degrees i and r. Furthermore it is startling that the dual of $H_I^i(T_i)$ is only one dimensional rather than i dimensional as might be expected: thus T_i is Cohen-Macaulay in codimension $r - 2$, and it turns out that T is Gorenstein in codimension $r - 2$.

From this information it follows that the ith local cohomology of R agrees with that of T_i except perhaps in dimensions $0, 1, 2$ and r. Indeed, there is only one connecting homomorphism between non-zero groups: $H_I^1(Q) \longrightarrow H_I^2(T_2)$. However for V of rank ≥ 3 there are non-trivial higher differentials. It is natural to view the above connecting homomorphism as d^1, and in fact the only nonzero differentials all originate in $H_I^1(Q)$, and each has the effect of replacing $H_I^1(Q)$ by twice the previous one, so that $E_{-1,*}^i = 2^{i-1} H^1(Q)$ for $1 \leq i \leq r$ and $E_{-1,*}^\infty = E_{-1,*}^r = 2^{r-1} H^1(Q)$. Furthermore the differentials $d^1, d^2, \ldots, d^{r-2}$ are all surjective: this can be deduced from the module structure together with the fact that $ku_*(BV)$ is connective. This means that the E^∞ term is concentrated on columns $0, -1$ and $-r$; the 0th column is the direct summand ku_*, and in fact we obtain a rather complete and natural description of the homology.

THEOREM 4.1.1. *The spectral sequence has E^∞-term on the columns $s = 0, -1$ and $-r$, and this gives an extension*

$$0 \longrightarrow \Sigma^{-4} T^\vee \longrightarrow \widetilde{ku}_*(BV) \longrightarrow \Sigma^{-1}(2^{r-1} H_I^1(Q)) \longrightarrow 0$$

of $GL(V)$-modules, and the extension is additively split. Here $T^\vee = \mathrm{Hom}(T, \mathbb{F}_2)$ is the $\mathbb{F}_2$-dual and the suspensions are such that the lowest nonzero group in the kernel is in degree 2 and the lowest nonzero group in the quotient is in degree 1.

The calculation is arranged as follows. In Section 4.2 the cohomology ring $ku^*(BV)$ is calculated by the method of Chapter 2. In Section 4.4 we calculate the local cohomology of Q, and deduce some facts about its 2-adic filtration in Section 4.5 for later use. The calculation of the local cohomology of T is more involved: in Section 4.6 we describe a free resolution and use it in Section 4.7 to calculate the local cohomology, with some Hilbert series calculations deferred to 4.8. In Section 4.10 we assemble the information to describe the local cohomology of R, and in Section 4.11 we find the differentials in the spectral sequence.

Finally, with a complete description of the homology and cohomology Sections 4.12 and 4.13 we discuss the homotopy Gorenstein duality statement of Section 3.3 and the Tate cohomology of Section 3.6. In this case the duality statement turns out to consist of the two isomorphisms

$$\mathrm{Ext}^2_{ku_*}(\Sigma^{-2} T^\vee, ku_*) = T$$

and

$$\mathrm{Ext}^1_{ku_*}(2^{r-1} H_I^1(Q), ku_*) = \widetilde{Q}$$

together with the fact that the other Ext groups vanish: the exact sequence from the universal coefficient theorem corresponds to that from the local cohomology theorem. The Tate cohomology displays both the extremely non-split behaviour of periodic K-theory [30], and the split behaviour of mod 2 cohomology [7].

4.2. The ku-cohomology of elementary abelian groups.

Ossa [48] calculated the connective ku cohomology of an elementary abelian group as a ku^*-module and described the ring structure of the image in periodic K-theory. Here, we reproduce his results and use the Adams spectral sequence to determine the ring structure, which greatly illuminates the additive structure. We will use the notation and results of 2.2.

We start with a characteristic free algebraic result. Let $E(1)$ be the exterior algebra over a field k of characteristic $p > 0$ on odd degree generators Q_0 and

Q_1. Let L be the infinite 'string module' (Figure 2.2) on a 0-dimensional class determined by the string $Q_0^{-1}Q_1Q_0^{-1}Q_1Q_0^{-1}Q_1\ldots$ (also called a 'lightning flash'). In detail, L has a k basis $\{x_{n,i} | i \in \{1,2\}, n \geq 0\}$ with operations $Q_0(x_{n,1}) = x_{n,2}$ and $Q_1(x_{n,1}) = x_{n+1,2}$. The class $x_{0,2}$ is the class from which all the others are reached by means of the string

$$Q_0^{-1}Q_1Q_0^{-1}Q_1Q_0^{-1}Q_1\ldots.$$

LEMMA 4.2.1. *With the diagonal $E(1)$ action on the tensor product,*

$$L \otimes L \cong L \oplus \bigoplus_{n,m \geq 0} <x_{n,1} \otimes x_{m,1}>$$

where the L on the right has initial class $x_{0,2} \otimes x_{0,2}$, and the rest is the free $E(1)$-module generated by the elements $x_{n,1} \otimes x_{m,1}$.

Proof: First note that we have at least a 2-adic integers worth of choices in the copy of L we use, since each Q_0^{-1} has two possible monomial values. For definiteness, let us choose $\{x_{0,2} \otimes x_{n,i}\}$. It is nontrivial but elementary to verify that the sum is direct and fills out all of $L \otimes L$. □

Since H^*BC_p is a sum of suspensions of L, as a module over $E(1)$, the preceding lemma is the key to understanding the Adams spectral sequence for the ku-cohomology of elementary abelian groups. We will use the following notation.

NOTATION 4.2.2. Let $B = BC_p$, and let $BV = B \times \cdots \times B$ be the classifying space of a rank r elementary p-group V. Let $\mathcal{N} = \{1, 2, \ldots, r\}$. Write $H^*(B_+) = E[x] \otimes \mathbb{F}_p[y]$ if $p > 2$, and $\mathbb{F}_2[x]$ if $p = 2$. In the latter case, let $y = x^2$. In H^*BV let x_i and y_i be the corresponding elements in the ith factor. We shall have occasion to consider H^*BV as a module over the subring $P = \mathbb{F}_p[y_1, y_2, \ldots, y_r]$. Write the complex representation rings $R(C_p) = \mathbb{Z}[\alpha]/(\alpha^p - 1)$ and $R(V) = \mathbb{Z}[\alpha_1, \ldots, \alpha_r]/(\alpha_1^p - 1, \ldots, \alpha_r^p - 1)$. For $S \subset \mathcal{N}$, let

$$V_S = \prod_{i \in S} V_i$$

where V_i is the ith factor C_p in V.

As a formal consequence of the stable equivalence $X \times Y \simeq X \vee Y \vee X \wedge Y$ we have a decomposition

$$BV_+ = \bigvee_{S \subset \mathcal{N}} B^{\wedge |S|}.$$

functorial for permutations of the basis.

As a consequence, stable invariants such as ku^*, K^* and H^* of BV split in the same way (as modules over the corresponding coefficients, but not as rings). There is a corresponding additive splitting of the complex representation ring

$$R(V) = \bigoplus_{S \subset \mathcal{N}} R_S$$

where R_S consists of those representations pulled back from V_S but no smaller quotient. That is, R_S is spanned by the

$$\prod_{i \in S} \alpha_i^{n_i}$$

with each $n_i \neq 0$ modulo p. At odd primes, we have the further splitting
(6) $$B = BC_p = B_1 \vee \cdots \vee B_{p-1},$$

4.2. THE ku-COHOMOLOGY OF ELEMENTARY ABELIAN GROUPS.

where B_i has cells in dimensions congruent to $2i-1$ and $2i$ modulo $2(p-1)$. Thus

$$B^{\wedge k} = \bigvee_{1 \leq i_j \leq p-1} B_{i_1} \wedge \cdots \wedge B_{i_k},$$

with concomitant splittings of ku^*, K^*, and H^* of $B^{\wedge k}$ and BV. The corresponding splitting of R_S is

$$\bigoplus_{1 \leq i_j \leq p-1} <\alpha_1^{i_1} \cdots \alpha_k^{i_k}>$$

when $S = \{1, \ldots, k\}$, and similarly for other S. We therefore have compatible splittings of ku^*, K^*, and H^* of BV, and a similar splitting of $R(V)$, all indexed on the basis $\{\alpha_1^{i_1} \cdots \alpha_r^{i_r} | 0 \leq i_j < p\}$ of $R(V)$. Therefore, to describe ku^*BV we shall compute each of the $ku^*(B_{i_1} \wedge \cdots \wedge B_{i_r})$ and the multiplication which ties them together.

Recall that the $ku \wedge B_i$ are all equivalent up to suspension (Lemma 2.2.2). For smash products of the B_i we have the following key consequence of Lemma 4.2.1, due to Ossa [48].

PROPOSITION 4.2.3. *For each* $I = (i_1, \ldots, i_k)$ *there is a generalized mod p Eilenberg-MacLane spectrum* X_I *and an equivalence*

$$ku \wedge B_{i_1} \wedge \cdots \wedge B_{i_k} \simeq X_I \vee ku \wedge B_i \simeq X_I \vee ku \wedge \Sigma^{2i-2} B_1$$

where $i = i_1 + \cdots + i_k$ *and we let* $B_{j+p-1} = \Sigma^{2(p-1)} B_j$.

Proof: By induction and Lemma 2.2.2 it is sufficient to prove the Proposition for $k = 2$ and $i_1 = i_2 = 1$. Since

$$H^*(ku \wedge B_1 \wedge B_1) \cong H^*ku \otimes H^*(B_1 \wedge B_1) \cong \mathcal{A} \underset{E(1)}{\otimes} H^*(B_1 \wedge B_1),$$

and $H^*(B_1 \wedge B_1) \cong \Sigma^2 L \otimes \Sigma^2 L \cong \Sigma^4(L \otimes L)$, the $E(1)$-free summand in Lemma 4.2.1 produces a corresponding $\mathcal{A}$-free summand in $H^*(ku \wedge B_1 \wedge B_1)$. Margolis' theorem [43] gives a correspondence between bases of $\mathcal{A}$-free submodules of $H^*(X)$ and $H\mathbb{F}_p$ wedge summands of X. This gives the generalized Eilenberg-MacLane wedge summand. Let C be the other summand. We wish to show that $C \simeq ku \wedge B_2$. The diagonal map $\Delta: B \longrightarrow B \wedge B$ induces the required equivalence

$$ku \wedge B_2 \longrightarrow ku \wedge B \xrightarrow{\Delta} ku \wedge B \wedge B \longrightarrow ku \wedge B_1 \wedge B_1 \longrightarrow C.$$

To see this, it suffices to check that, in cohomology, the composite

$$B_2 \longrightarrow B \longrightarrow B \wedge B \longrightarrow B_1 \wedge B_1$$

maps the summand $\Sigma^4 L$ from Lemma 4.2.1 isomorphically to $H^*(B_2)$. Since $\Sigma^4 L$ consists of $\{x_{0,2} \otimes x_{n,i}\}$, which maps to $\{y\} \otimes H^*(B_1)$ in $H^*(B \wedge B)$, Δ^* maps this to $H^*(B_2)$ as required. $\square$

We can now describe all the ku^*-module structure of $ku^*(BV)$, and most of its ring structure. The rank 2 case is shown in Figure 4.1. Note that in positive Adams filtration we have constant rank 3, as in $\widetilde{K}^*(BV_2)$, while in Adams filtration 0 we have polynomial growth as in $H^*(BV_2)$.

THEOREM 4.2.4.   (1) *The Adams spectral sequence*

$$\mathrm{Ext}_{\mathcal{A}}^{*,*}(H^*(ku), H^*(BV)) \Longrightarrow ku^*(BV)$$

collapses at E_2.

(2) There is a set of generators for $ku^*(BV)$ as a ku^*-module which is mapped monomorphically to a vector space basis by the edge homomorphism
$$ku^*(BV) \twoheadrightarrow \operatorname{Ext}^0(\mathbb{F}_p, H^*(BV)) = (Q_0, Q_1)\text{-}\operatorname{ann}(H^*(BV)).$$
(3) The edge homomorphism maps the subalgebra
$$ku^*(B) \underset{ku^*}{\otimes} \cdots \underset{ku^*}{\otimes} ku^*(B)$$
of $ku^*(BV)$ onto $P = \mathbb{F}_p[y_1, \ldots, y_r] \subset (Q_0, Q_1)\text{-}\operatorname{ann}(H^*(BV))$.

(4) As a ku^*-module, there is a splitting
$$ku^*(BV) \cong M \oplus ku^*(B) \underset{ku^*}{\otimes} \cdots \underset{ku^*}{\otimes} ku^*(B)$$
such that if $x \in M$ then then $px = 0 = vx$, and x^p is in the tensor product. The summand M is mapped monomorphically to $H^*(BV)$ by the edge homomorphism.

Proof: Any nonzero differential would have to occur on a finite skeleton, so its dual would occur in the Adams spectral sequence for $ku_*(BV)$. But this spectral sequence collapses by the geometric splitting 4.2.3 since this is so for both B and $H\mathbb{F}_p$.

Now (1) follows immediately from the collapse of the Adams spectral sequences for B and $H\mathbb{F}_p$, while (2) follows because it is true in both these Adams spectral sequences. Part (3) follows by naturality of the product.

The splitting in (4) also follows from the geometric splitting 4.2.3, since the tensor product summand contains all the suspensions of $ku \wedge B_1$ and some of the $H\mathbb{F}_p$'s. Thus the summand M is a sum of $\pi_*(H\mathbb{F}_p)$'s, so is annihilated by p and v and detected in Ext^0. Since the pth power of $(Q_0, Q_1)\text{-}ann(H^*(BV))$ is contained in $\mathbb{F}_p[y_1, \ldots, y_r]$, if $x \in M$ then x^p is detected by an element of $\mathbb{F}_p[y_1, \ldots, y_r]$. By (3), an element t of the tensor product subalgebra will be detected by this same element of $\mathbb{F}_p[y_1, \ldots, y_r]$, and hence $x^p - t$ will be detected in Adams filtration greater than 0. But all such elements are in the tensor product subalgebra, hence so is x^p. $\square$

REMARK 4.2.5. The last item in this theorem gives a direct proof of the central technical result of Chapter 1, Theorem 1.5.1 in this special case. This was the first indication that such a result might be true in general.

COROLLARY 4.2.6. *The map*
$$ku^*(B) \underset{ku^*}{\otimes} \cdots \underset{ku^*}{\otimes} ku^*(B) \longrightarrow ku^*(BV)$$
is a V-isomorphism.

We also have

COROLLARY 4.2.7. *The natural maps induce an injective ring homomorphism*
$$ku^*(BV) \longrightarrow K^*(BV) \times H^*(BV)$$

REMARK 4.2.8. This allows us to finish computing the multiplicative structure of $ku^*(BV)$ in principle. We shall have more to say about this shortly.

REMARK 4.2.9. Consider $p = 2$. The generators of the tensor product subalgebra satisfy
$$2y_i^2 y_j = vy_i^2 y_j^2 = 2y_i y_j^2 \quad \text{and} \quad vy_i^2 y_j = 2y_i y_j = vy_i y_j^2.$$

4.2. THE ku-COHOMOLOGY OF ELEMENTARY ABELIAN GROUPS.

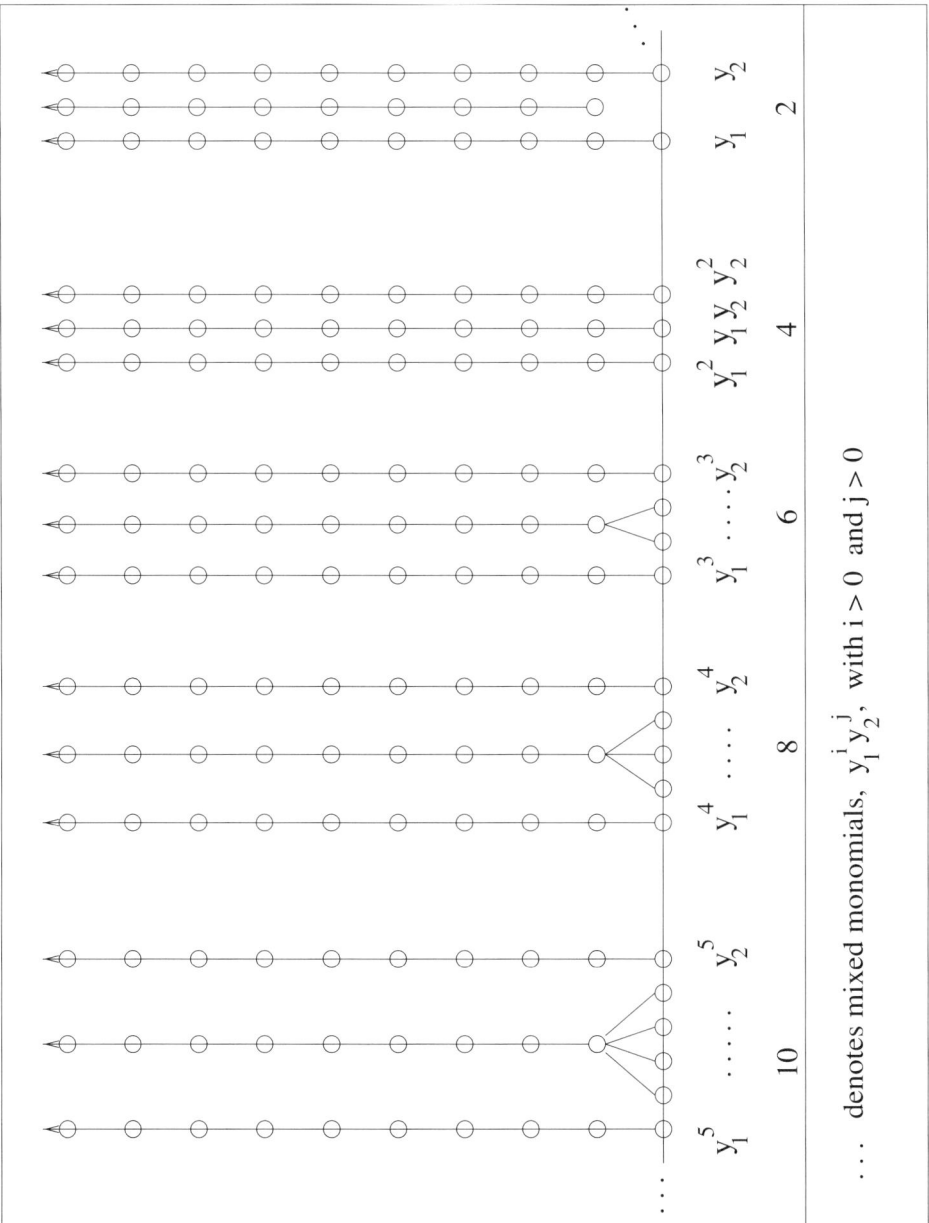

FIGURE 4.1. The $E_2 = E_\infty$ term of the Adams spectral sequence $\mathrm{Ext}_{E(1)}(\mathbb{F}_2, H^*(BV_2)) \Longrightarrow ku^*(BV_2)$

Since the v-torsion maps to 0 in the periodic K-theory of BV, we have $y_i^2 y_j = y_i y_j^2$ there. This rather neatly explains the 'somewhat weird structure' observed by Wall in his calculation of the image of cohomology in periodic K-theory in 1961 [**57**], in that it shows these relations follow immediately from the relation between 2 and v in the individual factors.

Here is a more explicit description of the subset of $H^*(BV)$ which detects our set of generators for $ku^*(BV)$. Recall that $P = \mathbb{F}_p[y_1, \ldots, y_r] \subset H^*(BV)$.

PROPOSITION 4.2.10.
$$(Q_0, Q_1) - ann(H^*(BV)) = P + \text{Im}(Q_0 Q_1).$$

Proof: Certainly the right hand side is (Q_0, Q_1)-annihilated. On the other hand, Lemma 4.2.1 implies that we need only consider the (Q_0, Q_1)-annihilated elements of the suspensions of L, which all lie in P, together with the (Q_0, Q_1)-annihilated elements of $E(1)$-free summands which are exactly the image $\text{Im}(Q_0 Q_1)$. □

We shall also need to know what is annihilated by the smaller ideal $(Q_1 Q_0)$.

PROPOSITION 4.2.11.
$$(Q_1 Q_0) - ann(H^*(BV)) = \text{Im}(Q_0) + \text{Im}(Q_1) + P + \bigoplus_i P x_i.$$

Proof: It is easy to check that the right hand side is annihilated by $Q_0 Q_1$. Let $N = \tilde{H}^*(BC_p^{\wedge r})$. It suffices to assume $r > 1$ and show $(Q_1 Q_0) - ann(N) = N'$, where
$$N' = Q_0(N) + Q_1(N) + P y_1 \cdots y_r + P y_1 \cdots y_{r-1} x_r.$$

Again it is clear that N' is annihilated by $Q_1 Q_0$.

By the Künneth formula, $H_*(N; Q_1)$ is the set of all $c y_1 \cdots y_r$ where $c \in P$ has degree less than $p-1$ in each variable y_j. Hence if $Q_1 Q_0 m = 0$ we deduce
$$Q_0 m = Q_1 n + c y_1 \cdots y_r,$$
where the dimension of n is smaller than the dimension of m. Applying Q_0 gives $Q_1 Q_0 n = 0$; by induction we conclude
$$Q_1 n \in \text{Im}(Q_1 Q_0) + P y_1 \cdots y_r.$$
Since $H_*(N; Q_0) = 0$ we obtain
$$m - c y_1 \cdots y_{r-1} x_r \in N'$$
and the result follows. □

REMARK 4.2.12. Again let $p = 2$. If $I = (i_1, \ldots, i_r)$ is a sequence of nonnegative integers, let $x^I = x_1^{i_1} \cdots x_r^{i_r}$ and let $2I = (2i_1, \ldots, 2i_r)$. If $S \subset \mathbb{N}$ let $x_S = \prod_{i \in S} x_i$. Then every monomial in the x_i has a unique expression as $x^{2I} x_S$ for some I and S. One computes that
$$Q_0 Q_1(x^{2I} x_S) = x^{2I} x_S \sum_{i,j \in S, i \neq j} x_i x_j^3.$$

Thus, as a module over $\mathbb{F}_2[x_1^2, \ldots, x_r^2] = \mathbb{F}_2[y_1, \ldots, y_r]$, the (Q_0, Q_1)-annihilated elements are generated by the 2^r elements $q_S = Q_0 Q_1(x_S)$. In fact, this is redundant, since $q_S \in \mathbb{F}_2[x_1^2, \ldots, x_r^2]$ if $|S| < 3$, so it suffices to use those q_S with $|S| \geq 3$. Thus, the module M defined in 4.2.4 is generated as a module over the tensor product subalgebra by $2^r - 1 - r - \binom{r}{2}$ elements detected by these q_S. A more precise analysis of the entire 2-torsion ($= v$-torsion) submodule can be found in Section 4.6.

For example, the first exotic generator (i.e., a class not in the tensor product subalgebra) occurs when V has rank 3 and is the element $q_{\{1,2,3\}} \in ku^7(BV) = \mathbb{Z}/2$. It is detected in Ext^0 by

$$Q_0 Q_1(x_1 x_2 x_3) = x_1 x_2^2 x_3^4 + x_1 x_2^4 x_3^2 + x_1^2 x_2 x_3^4 + x_1^2 x_2^4 x_3 + x_1^4 x_2 x_3^2 + x_1^4 x_2^2 x_3$$

By Corollary 4.2.7, the square $q_{\{1,2,3\}}^2 \in ku^{14}(BV)$ is the unique element whose image in periodic K-theory is 0, and which is detected by $(Q_0 Q_1(x_1 x_2 x_3))^2$ in Ext^0. The relations in $ku^*(BV)$ mean that this element has many representations, e.g.,

$$\begin{aligned} q_{\{1,2,3\}}^2 &= y_1 y_2^2 y_3^4 + y_1 y_2^4 y_3^2 + y_1^2 y_2 y_3^4 - y_1^2 y_2^4 y_3 - y_1^4 y_2 y_3^2 - y_1^4 y_2^2 y_3 \\ &= y_1 y_2^2 y_3^4 - y_1 y_2^4 y_3^2 + y_1^2 y_2 y_3^4 - y_1^2 y_2^4 y_3 - y_1^4 y_2 y_3^2 + y_1^4 y_2^2 y_3 \end{aligned}$$

This is $GL(V)$-invariant since $q_{\{1,2,3\}}$ is the unique nonzero element of $ku^7(BV)$, so is fixed by every automorphism. The referee points out that this can be written in a transparently invariant form:

$$q_{\{1,2,3\}}^2 = \det \begin{pmatrix} y_1 & y_2 & y_3 \\ y_1^2 & y_2^2 & y_3^2 \\ y_1^4 & y_2^4 & y_3^4 \end{pmatrix}.$$

4.3. What local cohomology ought to look like.

We now begin the process of calculating the local cohomology of the ring $R = ku^*(BV)$ when $p = 2$. This will occupy us until Section 4.11 and is principally a local cohomology calculation. We therefore spend this short section building appropriate expectations.

One of the attractions of our method is that our calculations do not involve choices. The choice of basis used in the preceding section to calculate the Adams spectral sequence for ku^*BV was a temporary expedient which has served its purpose. What this freedom from arbitrary choices means in practice is that one can identify additional structures. Most importantly, our answers are representations of $GL(V)$. It is also practical to track naturality for group homomorphisms $V \longrightarrow V'$, and to identify the action of cohomology operations.

In describing local cohomology we should compare with that of the best behaved modules. We describe the general behaviour of the I-local cohomology $H_I^* M$ of an R-module M, and then impose the conditions giving the best behaviour. First, the local cohomology modules vanish above the dimension d of the module M (i.e., above the Krull dimension of the ring $R/\mathrm{ann}(M)$). On the other hand the local cohomology $H_I^*(M)$ vanishes up to the I-depth of M, so that if there is an M-regular sequence of length l in I we find $H_I^i(M) = 0$ for $i < l$, and the potentially non-zero modules are

$$H_I^l(M), H_I^{l+1}(M), \ldots, H_I^d(M).$$

Thus if we restrict to modules which have I-depth equal to the dimension d (the so-called *Cohen-Macaulay* modules), the only non-vanishing local cohomology module is $H_I^d(M)$. The best case occurs when this has a duality property. We describe this in the simplest case when (R, I, k) is a local ring and R is a k-algebra. Here the duality property states

$$H_I^*(M) = H_I^d(M) = \Sigma^\nu M^\vee$$

for some ν, where $M^\vee = \operatorname{Hom}_k(M, k)$ is the k-dual of M. When this duality property holds, we say that M is a *Gorenstein* module of Krull dimension d. However we warn that there is usually no natural comparison map.

To give life to these remarks, we consider the polynomial ring $R = k[W]$ over a field k, where W is a vector space of dimension r, concentrated in degree -2 (i.e., codegree 2). Thus R is of Krull dimension r, and of depth r, so that $H_I^*(R)$ is concentrated in degree r. Choosing a basis of W it is easy to calculate $H_I^r(R)$ directly, and hence to check that R is Gorenstein with $\nu = -2r$. More concretely, we may take $r = 1$ and consider the ring $R = k[x]$ with x of degree -2. Thus $H_I^*(k[x]) = H_I^1(k[x]) = k[x, x^{-1}]/k[x]$. This has dual $\Sigma^{-2} k[x]$ so that R is Gorenstein. However if a cyclic group $C = \langle g \rangle$ acts on $k[x]$ via $g \cdot x = \lambda x$ for some scalar λ, then g acts on the bottom element of $H_I^1(R)$ as $1/\lambda$. The corresponding fact for an arbitrary representation W of dimension r is the equivariant isomorphism

$$H_I^r(k[W])^\vee = \Sigma^{-2r} k[W] \otimes det(W).$$

The case that most concerns us is $k = \mathbb{F}_2$, in which case the determinant is necessarily trivial.

Even though not all modules are Cohen-Macaulay, the higher local cohomology modules are always the most significant. More precisely, if M is finitely generated, $H_I^i(M)$ is an Artinian module so it is natural to consider its dual $H_I^i(M)^\vee$ which is a Noetherian module when R is local. It turns out that $H_I^i(M)^\vee$ is of dimension $\leq i$. In our case this has very concrete implications. Indeed, we will be working over $R = k[W]$ with graded modules of finite dimension over a field in each degree; in that case, being of dimension $\leq i$ means that the Hilbert series is bounded by a polynomial of degree $i - 1$. We will find that the local cohomology modules that arise in our examples deviate from the Gorenstein condition by much less than this generic amount: the difference will consist of modules whose dual is only one dimensional.

We pause to formulate some notation for suspensions. Firstly, we often use the algebraists' notation

$$M(n) = \Sigma^n M$$

for typographical reasons. We also need to deal with modules M that are zero above a certain degree; the dual $M^\vee$ of such a module will therefore be zero below a certain degree. It is convenient to use the notation

$$\operatorname{Start}(i) M^\vee$$

for the suspension of $M^\vee$ whose lowest nonzero degree is i. The notation $\operatorname{Start}(i) N$ always implies that N is a non-zero bounded below module.

4.4. The local cohomology of Q.

The module Q is the image of $ku^*(BV)$ in $K^*(BV)$, and is therefore the Rees ring of the completed representation ring for the augmentation ideal. It is thus generated by $1, v, y_1, \ldots, y_r$; the elements y_i are the images of the ku-Euler classes, and therefore $vy_i = 1 - \alpha_i$.

We begin by giving a description of Q as an abelian group: this is the only part of the calculation in which we are working over the integers rather than over $\mathbb{F}_2$. First, Q is generated by monomials $v^s y^I$ where $I = (i_1, \ldots, i_r)$. Next, note that because we have factored out T, a monomial y^I is determined by its degree and the

4.4. THE LOCAL COHOMOLOGY OF Q.

subset $\mathrm{Supp}(I) = \{s \mid i_s \neq 0\}$ of indices which occur in it. Therefore, if $S \neq \emptyset$, we write y_S^n for the element of degree $-2n$ in which the set of indices which occur is S. We also permit $y_\emptyset = 1$, although the empty set is exceptional at several points. By construction $y_i y_S^n = y_{S \cup \{i\}}^{n+1}$. Finally $v y_S^{n+1} = 2 y_S^n$ provided y_S^n exists (i.e., provided the subset S has $\leq n$ elements).

LEMMA 4.4.1. *As an abelian group, Q is the direct sum of $\mathbb{Z}[v]$ and a free module over $\mathbb{Z}_2^\wedge$. The free module has basis the monomials y_S^n with $n \geq |S|$ and $v^i y_S^{|S|}$ for $i \geq 0$.* □

REMARK 4.4.2. It is helpful to display Q in the style of an Adams spectral sequence. We filter it by powers of the ideal $(2, v)$ and display the associated graded module, with the subquotients of Q_i in a column above i, each basis element of the resulting graded vector space represented by a dot, and vertical lines representing multiplication by 2. Thus if $n \geq |S|$, the element y_S^n contributes a dot at $(-2n, 0)$ and $2^i v^j y_S^n$ contributes a dot at $(-2n + 2j, i + j)$.

Although we want $H_I^*(Q)$ as a module over ku^*, the calculation of the abelian group only requires the action of $P\mathbb{Z} = \mathbb{Z}[y_1, y_2, \ldots, y_r]$, and it is convenient to let Q' be the $P\mathbb{Z}$-submodule of elements of degree $\leq -2r$. Thus as an abelian group, Q' is free over $\mathbb{Z}_2^\wedge$ of rank $2^r - 1$ in each even degree $\leq -2r$. We define Q'' by the exact sequence

$$0 \longrightarrow Q' \longrightarrow Q \longrightarrow Q'' \longrightarrow 0.$$

Since Q'' is bounded below, it is I-power torsion and

$$H_I^*(Q'') = H_I^0(Q'') = Q''.$$

Now let us turn to the module Q'. We place great value on making our answers functorial in V, and in particular they should be equivariant for $GL(V)$. The action on Q can be made explicit from the matrix form of an element of $GL(V)$ by using formal addition in place of conventional addition. For example, if V is of rank 2, the matrix

$$\begin{pmatrix} 0 & 1 \\ 1 & 1 \end{pmatrix}$$

takes

$$y_2 \longmapsto y_1 \odot y_2 = y_1 + y_2 - v y_1 y_2 = y_1 + y_2 - v y_{12}^2.$$

Notice that the matrix entries are mod 2, whilst Q is torsion free. Since $[2](y_i) = y_i \odot y_i = y_i + y_i - v y_i^2 = y_i + y_i - 2 y_i = 0$, this is legitimate.

Evidently it is very useful to pick out an invariant element. We do this by working from the obviously invariant element

$$\begin{aligned} \rho &= \sum_{\alpha \in V^\vee} \alpha \\ &= |V| - \sum_{\alpha \in V^\vee} v e_{ku}(\alpha) \\ &= \sum_{i=0}^r 2^{r-i} (-v)^i \sum_{|S|=i} y_S^i. \end{aligned}$$

LEMMA 4.4.3. *The module Q contains a unique element $y^* \in Q_{-2r}$ with the property*

$$v^r y^* = |V| - \rho.$$

and this element is invariant under $GL(V)$.

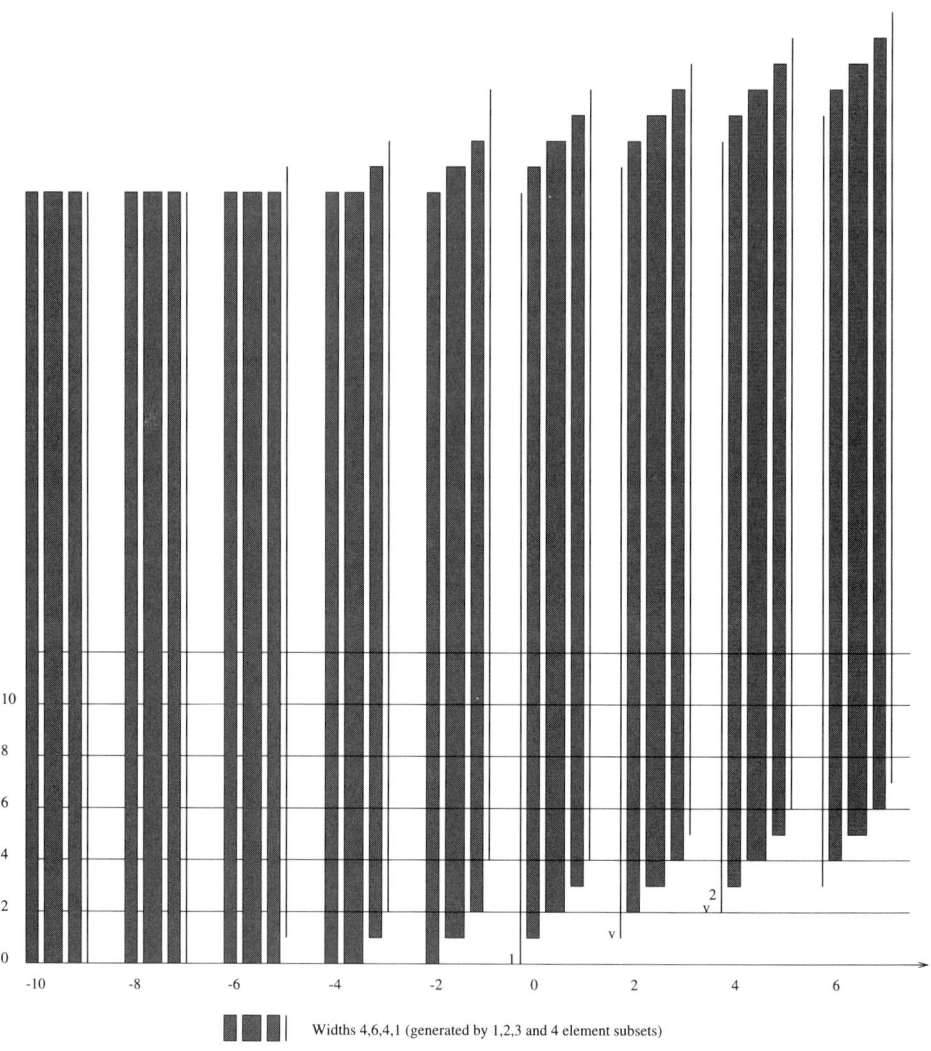

FIGURE 4.2. The module Q in rank 4.

Proof: Since v is a monomorphism on Q, there is at most one such element y^*, and since ρ is $GL(V)$-invariant so is y^*. To show y^* exists we note that

$$y^* = \sum_{S \neq \emptyset} (-1)^{|S|} y_S^r$$

has the required property. □

LEMMA 4.4.4. *The element y^* is Q'-regular and furthermore $y^* : Q'_a \longrightarrow Q'_{a-2r}$ is an isomorphism when $Q'_a \neq 0$.*

Proof: It suffices to show $v^r y^*$ is regular. However the y_i's all annihilate ρ, so $v^r y^*$ acts as $|V|$. Since Q' has no $\mathbb{Z}$-torsion, y^* is regular. Since v^r acts as multiplication by $|V|$ we see that multiplication by y^* is also surjective. □

4.4. THE LOCAL COHOMOLOGY OF Q.

LEMMA 4.4.5. *The module Q' is of dimension 1 and*
$$H_I^*(Q') = H_I^1(Q') = Q'[1/y^*]/Q'.$$

Proof: We may begin by calculating local cohomology for the principal ideal $(y^*) \subseteq I$. From 4.4.4 we see
$$H_{(y^*)}^*(Q') = H_{(y^*)}^1(Q') = Q'[1/y^*]/Q'.$$
This is in particular bounded below, so I-power torsion, and hence equal to $H_I^*(Q')$. □

REMARK 4.4.6. It seems that the module Q' is also Gorenstein in the sense that
$$H_I^*(Q') = H_I^1(Q') = \mathrm{Start}(-2r+2)(Q')^\vee.$$
To establish this we might use the product
$$Q[1/y^*] \otimes Q[1/y^*] \longrightarrow Q[1/y^*].$$
We then notice that the periodicity y^* (of degree $-2r$) has an rth root, $2/v$ (of degree -2). This allows us to pair complementary degrees to $Q[1/y^*]_0$. Finally, we need to show that this admits a map to $\mathbb{Z}_2^\wedge$ giving a perfect pairing.

So far we have only checked this in rank 2. In this case the map is
$$\lambda_1 y_1^n + \lambda_2 y_2^n + \lambda_{12} y_{12}^n \longmapsto 2(\lambda_1 + \lambda_2) + \lambda_{12},$$
and the duality gives dual bases
$$\{y_1^{-n}, y_2^{-n}, y_{12}^{-n}\} \text{ and } \{y_1^n - y_{12}^n, y_2^n - y_{12}^n, -y_1^n - y_2^n + 3y_{12}^n\}.$$
There is an obvious generalization of the augmentation to the general case, but to see it is $GL(V)$-invariant and gives a duality, we need to describe it neatly in terms of representation theory. But be warned that the image in periodic K-theory is *not* the augmentation ideal, for which this construction does not work.

PROPOSITION 4.4.7. *The local cohomology of Q is given by*
$$H_I^i(Q) = \begin{cases} ku^* \cdot \rho & \text{if } i = 0 \\ Q[1/y^*]/Q & \text{if } i = 1 \\ 0 & \text{otherwise} \end{cases}$$
For practical purposes, $H_I^1(Q)$ is best calculated by the exact sequence
$$0 \longrightarrow ku^* \cdot \rho \longrightarrow Q'' \longrightarrow Q'[1/y^*]/Q' \longrightarrow H_I^1(Q) \longrightarrow 0.$$
We may make explicit what this is in degree $-2r + 2n$. In negative degrees we have
$$H_I^1(Q)_{-2r+2n} = \mathbb{Z}_2^\wedge\{y_S^{2n-2r} \mid S \neq \emptyset\}/(2^{|S|-r+n} y_S^{2n-2r})$$
where negative powers of 2 are treated as 1. As an abelian group this is
$$\binom{r}{r}\mathbb{Z}/2^n \oplus \binom{r}{r-1}\mathbb{Z}/2^{n-1} \oplus \binom{r}{r-2}\mathbb{Z}/2^{n-2} \oplus \cdots \oplus \binom{r}{1}\mathbb{Z}/2^{n-r+1}$$
where $\binom{r}{i}$ is a binomial coefficient. In degree zero and above we have
$$H_I^1(Q)_{-2r+2n} = \mathbb{Z}_2^\wedge\{y_S^{2n-2r} \mid S \neq \emptyset\}/(2^{|S|-r+n} y_S^{2n-2r}, v^{n-r})$$

92   4. ELEMENTARY ABELIAN GROUPS

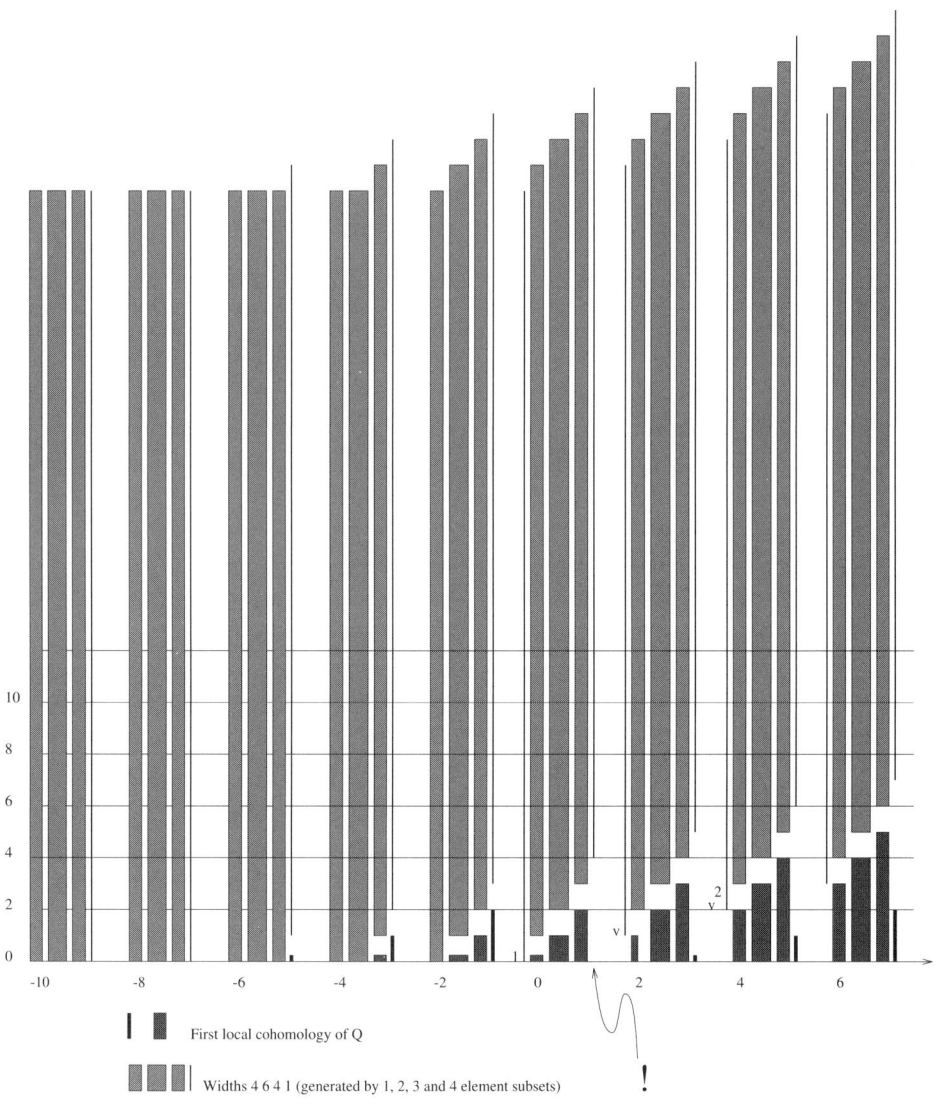

FIGURE 4.3. The first local cohomology of Q in rank 4.

where
$$v^{n-r} = 2^{n-r} \cdot \left(\sum_{S \neq \emptyset} (-1)^{|S|} y_S^{2n-2r} \right).$$

As an abelian group this is

$$\binom{r}{r}\mathbb{Z}/2^{n-r} \oplus \binom{r}{r-1}\mathbb{Z}/2^{n-1} \oplus \binom{r}{r-2}\mathbb{Z}/2^{n-2} \oplus \cdots \oplus \binom{r}{1}\mathbb{Z}/2^{n-r+1}.$$

Proof: First consider $H_I^0(Q)$. By definition Q has no v torsion, and since T is v-power torsion, Q is a submodule of $ku^*(BV)[1/v] = K^*(BV)$. Hence $H_I^0(Q)$ is the submodule of $H^0_I(K^*(BV)) = K^* \cdot \rho$ consisting of elements from $ku^*(BV)$. This calculates $H^0_I(Q) = ku^* \cdot \rho$.

Consider the exact sequence
$$0 \longrightarrow Q' \longrightarrow Q \longrightarrow Q'' \longrightarrow 0.$$
This shows that Q is of dimension 1, and gives an exact sequence
$$0 \longrightarrow ku^* \cdot \rho \longrightarrow Q'' \longrightarrow H^1_I(Q') \longrightarrow H^1_I(Q) \longrightarrow 0.$$
The image of v^{n-r} is identified from the fact that $|V| - \rho = v^r y^*$ and the fact that ρ is trivial in $Q[1/y^*]$. □

4.5. The 2-adic filtration of the local cohomology of Q.

In the analysis of the spectral sequence we need a number of facts about the subquotients $2^{i-2}H^1_I(Q)/2^{i-1}H^1_I(Q)$, and we prove them here. We need their Hilbert series and an estimate for the degrees of generators.

LEMMA 4.5.1. *For $i = 2, 3, \ldots, r$, the graded $\mathbb{F}_2$-vector space*
$$(2^{i-2}H^1_I(Q)/2^{i-1}H^1_I(Q))^\vee$$
has Hilbert series $t^{2r-2i+2}((1+x)^r - x^{r-i+1})/(1-x)$, where $x = t^2$.

Proof: In 4.4.7 we calculated $H^1_I(Q)$ as an abelian group. We immediately deduce the subgroup $2^{i-2}H^1_I(Q)$ in degree $-2r+2n$. Note first that $H^1_I(Q)$ has an element of order 2^{r-1}, so that in all cases $H^1_I(Q)$ is non-zero in some negative degree. We write the group as a sum of cyclic groups $\mathbb{Z}/2^j$ with the convention that if $j \leq 0$ the contribution is zero. In negative degrees it is
$$\binom{r}{r}\mathbb{Z}/2^{n-i+2} \oplus \binom{r}{r-1}\mathbb{Z}/2^{n-i+1} \oplus \binom{r}{r-2}\mathbb{Z}/2^{n-i} \oplus \cdots \oplus \binom{r}{1}\mathbb{Z}/2^{n-r-i+3}$$
In degree zero and above it is
$$\binom{r}{r}\mathbb{Z}/2^{n-r-i+2} \oplus \binom{r}{r-1}\mathbb{Z}/2^{n-i+1} \oplus \binom{r}{r-2}\mathbb{Z}/2^{n-i} \oplus \cdots \oplus \binom{r}{1}\mathbb{Z}/2^{n-r-i+3}.$$
Now consider the dimension of the graded vector space $2^{i-2}H^1_I(Q)/2^{i-1}H^1_I(Q))$ as the degree increases through the even numbers. It is zero until degree $-2r+2i-2$ (i.e., $n = i-1$), when it adds the binomial coefficients $\binom{r}{r}, \binom{r}{r-1}, \ldots, \binom{r}{0}$ in turn, except that as we pass zero the number added is one less. This corresponds exactly to the calculation of $((1+x)^r - x^{r-i+1})(1 + x + x^2 + x^4 + \cdots)$. □

The quotients of the 2-adic filtration of the local cohomology of Q are naturally modules over $P = \mathbb{F}_2[y_1, y_2, \ldots, y_r]$.

LEMMA 4.5.2. *The P-module $(2^{i-2}H^1_I(Q)/2^{i-1}H^1_I(Q))^\vee$ is generated by its elements in degrees ≥ 0.*

Proof: Let $H = 2^{i-2}H^1_I(Q)/2^{i-1}H^1_I(Q)$. This is a module concentrated in even degrees $\geq -2r+2i-2$. The statement to be proved is equivalent to showing that if $j > 0$ then any map $\eta : H(-2j) \longrightarrow \mathbb{F}_2$ factors as $H(-2j) \longrightarrow H \longrightarrow \mathbb{F}_2$ where the first map is multiplication by an element of P.

On the other hand, our calculation of $H^1_I(Q)$ as a quotient of $Q[1/y^*]$ shows that a basis of H_{2j} is given by the images of $2^{i-2}y_S$ for $i - j - 1 \leq |S| \leq r - 1$ and

$|S| = r$ if $j \geq i - 1$. It suffices to deal with maps $\eta = \eta_S^j$ from the dual basis, and we do this by induction on j.

Suppose then that the elements η_S^j are all realized, and consider the problem of realizing η_S^{j+1}. There are three cases according to which subsets S qualify to give generators for j and $j+1$.

If the same subsets qualify as generators for j and $j+1$ (i.e., if $j \geq i-1$), we argue by induction on $|S|$. Suppose that $\eta_{|T|}^{j+1}$ is realized if $|T| < |S|$; this is vacuously true if $|S| = 1$. Now choose $s \in S$ and notice that

$$\eta_S^j y_s = \eta_S^{j+1} + \eta_{S\setminus\{s\}}^{j+1}.$$

In the other cases $2^{s-2} y_{\{1,2,\ldots,r\}}$ does not qualify as a basis element and by 4.4.7 is equal to the sum of all basis elements.

The next case is if the new subset qualifying for $j+1$ is $\{1,2,\ldots,r\}$ (i.e., if $j = i-2$). Then we note that all proper subsets qualify for j. Note first that

$$\eta_{\{1,2,\ldots,r\}\setminus\{s\}}^j y_s = \eta_{\{1,2,\ldots,r\}}^{j+1} + \eta_{\{1,2,\ldots,r\}\setminus\{s\}}^{j+1}.$$

Now argue by induction on $|S|$, assuming η_T^{j+1} is realized for $|T| < |S| \leq r$; this is vacuously true if $|S| = 1$. Now choose $s \in S$ and note that

$$\eta_S^j y_s = \eta_S^{j+1} + \eta_{S\setminus\{s\}}^{j+1} + \eta_{\{1,2,\ldots,r\}}^{j+1} + \eta_{\{1,2,\ldots,r\}\setminus\{s\}}^{j+1}.$$

The final case is if the new subsets qualifying for $j+1$ are the smallest (ie if $0 \leq j \leq i-3$). To start with, if $|S| = r-1$ we let $t \notin S$ be the omitted number and then

$$\eta_S^j y_t = \eta_S^{j+1}.$$

Now we argue by downwards induction on $|S|$, supposing that η_T^{j+1} is realized for $r > |T| > |S|$. Again we choose $t \notin S$ and note that

$$\eta_{S\cup\{t\}}^j y_t = \eta_S^{j+1} + \eta_{\{1,2,\ldots,r\}\setminus\{t\}}^{j+1} + \eta_{S\cup\{t\}}^{j+1}.$$

This completes the proof. □

REMARK 4.5.3. The critical use of the relation for $y_{\{1,2,\ldots,r\}}$ shows that the proof does not extend to showing $H^\vee$ is generated in strictly positive degrees.

4.6. A free resolution of T.

We now turn to the $(2,v)$-power torsion submodule T. We first recall that T is exact $(2,v)$-torsion. Thus all calculations in this section are mod 2, and indeed over the polynomial ring $P = \mathbb{F}_2[y_1, y_2, \ldots, y_r]$. The action of $GL(V)$ on P is now standard. The action of $GL(V)$ on local cohomology has not been made explicit. However since the discussion is in terms of the polynomial ring, it can be read off in a straightforward way.

The structure now becomes more intricate. We recall that we may view the polynomial ring $PE = \mathbb{F}_2[x_1, x_2, \ldots, x_r]$ in elements x_i of degree 1 as a module over $P = \mathbb{F}_2[y_1, y_2, \ldots, y_r]$ where y_i acts via x_i^2. As such $PE \cong P \otimes E$ where E is an exterior algebra on $x_1, x_2, \ldots, x_r$, and PE is a free P-module on the square-free monomials in $x_1, x_2, \ldots, x_r$.

4.6. A FREE RESOLUTION OF T.

Now T is the P-submodule of PE generated by certain elements q_S of degree $-|S|-4$, where S is a subset of $\{1,2,\ldots,r\}$ with at least 2 elements. By definition

$$q_S = Q_1 Q_0 x_S \text{ where } x_S = \prod_{s \in S} x_s.$$

One calculates that

$$q_S = x_S \sum_{s,t \in S}(x_s x_t^3 + x_s^3 x_t) = \sum_{s,t \in S} x_{S \setminus \{s,t\}}(x_s^2 x_t^4 + x_s^4 x_t^2) = \sum_{s,t \in S} x_{S \setminus \{s,t\}}(y_s y_t^2 + y_s^2 y_t).$$

Now consider the submodules

$$T_i = (q_S \mid |S| = i).$$

Since every monomial in the x's in q_S has exactly $|S|-2$ odd exponents, it follows that T_i lies in the free P submodule of PE on the degree $i-2$ part of E. Hence in particular, the submodules T_i have trivial intersection, and

$$T = \bigoplus_{i=2}^{r} T_i.$$

We now set about constructing a free P-resolution of T_i: it will turn out that a suitable truncation of a double Koszul complex will give the presentation.

We will need to consider modules $\binom{r}{j} P(-k)$, where $\binom{r}{j}$ is the binomial coefficient counting j-element subsets of a set with r elements. We use the basis $\{x_k(S)\}_S$ where S runs through subsets of $\{1,2,\ldots,r\}$ with exactly j elements. The subscript gives the degree.

LEMMA 4.6.1. *There is a presentation*

$$\binom{r}{i+1} P(-i-6) \oplus \binom{r}{i+1} P(-i-8) \xrightarrow{\langle d_0, d_1 \rangle} \binom{r}{i} P(-i-4) \longrightarrow T_i \longrightarrow 0,$$

where

$$d_0(x_{i+6}(T)) = \sum_{t \in T} y_t x_{i+4}(T \setminus \{t\})$$

and

$$d_1(x_{i+8}(T)) = \sum_{t \in T} y_t^2 x_{i+4}(T \setminus \{t\})$$

REMARK 4.6.2. This immediately shows that T_r is free on a single generator of degree $-r-4$. All other modules require more detailed analysis.

Proof: By definition T_i is generated over P by $\binom{r}{i}$ generators of degree $-i-4$. This establishes exactness at T_i.

The key to exactness at the next stage is that the action of $Q_1 Q_0$ is P-linear, together with the fact (4.2.11) that

$$\ker(Q_1 Q_0) = \operatorname{im}(Q_1) + \operatorname{im}(Q_0) + P + \bigoplus_{i=1}^{r} Px_i.$$

First we note that the composite at $\binom{r}{i} P(-i-4)$ is zero. We calculate

$$\begin{aligned}
\epsilon d_0 x_{i+6}(T) &= \epsilon \sum_{t \in T} y_t x_{i+4}(T \setminus \{t\}) \\
&= \sum_{t \in T} y_t q_{T \setminus \{t\}} \\
&= Q_1 Q_0 \sum_{s \in S} y_t x_{T \setminus \{t\}} \\
&= Q_1 Q_0 (Q_0 x_T) \\
&= 0.
\end{aligned}$$

Similarly $\epsilon d_1 x_{i+8}(T) = Q_1 Q_0 (Q_1 x_T) = 0$.

Now suppose
$$\epsilon(\sum_S \lambda_S x_{i+4}(S)) = \sum_S \lambda_S q_S = 0.$$

By P-linearity of $Q_1 Q_0$, the last equality is of the form
$$Q_1 Q_0 (\sum_S \lambda_S x_S) = 0.$$

Now the submodule of PE annihilated by $Q_1 Q_0$ is the sum of $P + \oplus P x_i$ and the images of Q_1 and of Q_0. Since $\sum \lambda_S x_S$ lies in the part of PE spanned by monomials with $|S|$ odd exponents, and $|S| = i \geq 2$, we see that
$$\sum_S \lambda_S x_S = Q_1 z_1 + Q_0 z_0.$$

Since
$$Q_0 x_T = \sum_t y_t x_{T \setminus \{t\}},$$
if we write $z_0 = \sum_T \mu_T^0 x_T$, and similarly $z_1 = \sum_T \mu_T^1 x_T$ with the sums over T with $|T| = i + 1$, then
$$\lambda_S = \sum_{T = S \cup \{t\}} y_t \mu_T^0 + y_t^2 \mu_T^1$$
and
$$d(\sum_T \mu_T^0 x_{i+6}(T) + \mu_T^1 x_{i+8}(T)) = \sum_S \lambda_S x_{i+4}(S)$$
as required. □

Now that we have the start of a resolution visibly related to the Koszul complexes for the regular sequences $y_1, y_2, \ldots, y_r$ and $y_1^2, y_2^2, \ldots, y_r^2$ it is not hard to continue it.

PROPOSITION 4.6.3. *The module T_i has a free resolution*
$$0 \longrightarrow F_{r-i} \longrightarrow F_{r-i-1} \longrightarrow \cdots \longrightarrow F_0 \longrightarrow T_i \longrightarrow 0,$$
where
$$F_s = \binom{r}{i+s} [P(-i - 4 - 2s) \oplus P(-i - 6 - 2s) \oplus \cdots \oplus P(-i - 4 - 4s))].$$
Note that this involves generators $x_j(S)$ where S has $i + s$ elements and
- $j \equiv i \mod 2$ *and*
- $2|S| < j < 4|S|$

The differential is
$$d(x_j(S)) = \sum_{s \in S} y_s x_{j-2}(S \setminus \{s\}) + y_s^2 x_{j-4}(S \setminus \{s\}),$$
where $x_j(T)$ is interpreted as zero unless the two displayed conditions are satisfied.

REMARK 4.6.4. It follows that the rank of T_i is $\binom{r-2}{i-2}$.

4.6. A FREE RESOLUTION OF T.   97

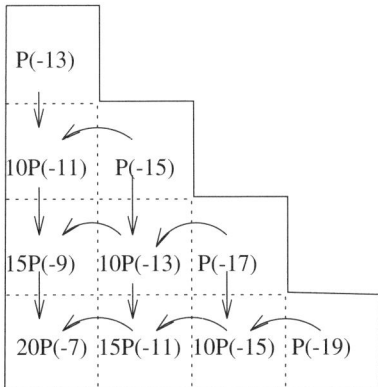

FIGURE 4.4. The double complex resolution for T_3 in rank 6.

Proof: We construct the above resolution as a truncation of an exact complex. For definiteness, we give the argument for T_{2k} in detail, and explain the modifications for T_j with j odd.

Indeed, we form the double Koszul complex K as the free P-module on generators $y_k(S)$ of internal degree $-2k$. We may define two differentials

$$d_0 y_k(S) = \sum_{s \in S} y_s y_{k-1}(S \setminus \{s\})$$

and

$$d_1 y_k(S) = \sum_{s \in S} y_s^2 y_{k-2}(S \setminus \{s\}).$$

It is easy to verify that $d_0^2 = 0$, $d_1^2 = 0$ and $d_0 d_1 = d_1 d_0$. We thus obtain a differential $d = d_0 + d_1$. Now the homological degree of $y_k(S)$ is $|S|$, and it is convenient to display K with $y_k(S)$ at $(k - |S|, 2|S| - k)$. This means that d_0 moves down one step and d_1 moves left one step. This suggests introducing a filtration by left half-planes:

$$\cdots \subseteq K_p \subseteq K_{p+1} \subseteq K_{p+2} \subseteq \cdots \subseteq K$$

where

$$K_p = \langle y_k(S) \mid k - |S| \leq p \rangle.$$

This gives rise to a spectral sequence

$$E_{p,q}^0 = H_{p,q}(K_p, K_{p-1}),$$

standard in the homological grading, so that the differentials d_0 and d_1 defined above are named so as to fit the standard spectral sequence notation.

Note that by construction K_p/K_{p-1} is the Koszul complex for the sequence $y_1, y_2, \ldots, y_r$. Accordingly, since $y_1, y_2, \ldots, y_r$ is a regular sequence in P, it follows that d_0 is exact except in the bottom nonzero degree in each column. Since this is in homological degree 0, there are no other differentials. We conclude that (K, d) is exact except in homological degree 0.

Now the proposed resolution $S = S(T_{2k})$ of T_{2k} is the quotient complex of K represented in the plane by the first quadrant with bottom corner generated by

$y_{k+2}(S)$ with $|S| = 2k$ (i.e., at $(2-k, 3k-2)$). From 4.6.1 we know that the bottom homology of S is T_{2k}, and it remains to show that S is exact except at the bottom. We deduce this from acyclicity of K. Indeed, since S_p/S_{p-1} is a truncation of a Koszul complex, $E_1(S)$ is a chain complex C concentrated at the bottom edge (i.e., $q = 3k - 2$), and a diagram chase establishes that d_1 is exact on C except at the bottom. Suppose $x \in C$ is a d_1-cycle not in the bottom degree; we show that x is a d_1-boundary. By definition of C, $x = [\hat{x}]$ for some $\hat{x}$, where $[\cdot]$ denotes d_0-homology classes. Since x is a cycle, there is $\hat{z}$ so that $d_1\hat{x} = \hat{y}$ and $d_0\hat{z} = \hat{y}$. Now $d(\hat{x} + \hat{z}) = d_0\hat{x} + d_1\hat{z}$. Since $d_0\hat{x}$ and $d_1 z$ are d-cycles, and we are above homological degree 0, there are elements s, u with $ds = \hat{x}$ and $du = \hat{z}$, and we find $d(\hat{x} + \hat{z} + s + u) = 0$. Hence there is a v with $dv = \hat{x} + \hat{z} + s + u$. Resolving v into its components we find $\hat{x} = d_0 v' + d_1 v''$ and so

$$x = [\hat{x}] = [\hat{x} + d_0 v'] = [d_1 v''] = d_1 [v'']$$

as required.

Finally, note that the only fact about P used above was that $y_1, y_2, \ldots, y_r$ is a P-regular sequence. We may therefore replace P by the odd degree part of PE and obtain the desired conclusion for T_j with j odd. □

EXAMPLE 4.6.5. (i) If $r = 2$ then $T = T_2$ is free of rank 1 over P on a generator of degree -6.
(ii) If $r = 3$ then $T = T_2 \oplus T_3$ where T_3 is free of rank 1 over P on a generator of degree -7, and T_2 admits a resolution

$$0 \longleftarrow T_2 \longleftarrow \binom{3}{2} P(-6) \longleftarrow \binom{3}{3}[P(-8) \oplus P(-10)] \longleftarrow 0.$$

(iii) If $r = 4$ then $T = T_2 \oplus T_3 \oplus T_4$ where T_4 is free of rank 1 on a generator of degree -8, T_3 has presentation

$$0 \longleftarrow T_3 \longleftarrow \binom{4}{3} P(-7) \longleftarrow \binom{4}{4}[P(-9) \oplus P(-11)] \longleftarrow 0$$

and T_2 has a presentation

$$0 \longleftarrow T_2 \longleftarrow \binom{4}{2} P(-6) \longleftarrow \binom{4}{3}[P(-8) \oplus P(-10)]$$
$$\longleftarrow \binom{4}{4}[P(-10) \oplus P(-12) \oplus P(-14)] \longleftarrow 0.$$

(iv) If $r = 5$ then $T = T_2 \oplus T_3 \oplus T_4 \oplus T_5$ where T_5 is free of rank 1 on a generator of degree -9, T_4 has a presentation

$$0 \longleftarrow T_4 \longleftarrow \binom{5}{4} P(-8) \longleftarrow \binom{5}{5}[P(-10) \oplus P(-12)] \longleftarrow 0,$$

T_3 has a resolution

$$0 \longleftarrow T_3 \longleftarrow \binom{5}{3} P(-7) \longleftarrow \binom{5}{4}[P(-9) \oplus P(-11)]$$
$$\longleftarrow \binom{5}{5}[P(-11) \oplus P(-13) \oplus P(-15)] \longleftarrow 0.$$

and T_2 has a resolution

$$0 \longleftarrow T_2 \longleftarrow \binom{5}{2}P(-6) \longleftarrow \binom{5}{3}[P(-8) \oplus P(-10)]$$

$$\longleftarrow \binom{5}{4}[P(-10) \oplus P(-12) \oplus P(-14)]$$

$$\longleftarrow \binom{5}{5}[P(-12) \oplus P(-14) \oplus P(-16) \oplus P(-18)] \longleftarrow 0.$$

4.7. The local cohomology of T.

Now that we have detailed homological control over T, we may calculate its local cohomology. Quite generally, if we have a free resolution

$$0 \longrightarrow F_r \longrightarrow F_{r-1} \longrightarrow \cdots \longrightarrow F_0 \longrightarrow M \longrightarrow 0$$

this gives a means of calculating $H_I^*(M)$; this is really just local duality, but it is helpful to make it explicit. Indeed, we tensor the resolution with the stable Koszul complex and obtain a double complex. Both the resulting spectral sequences collapse (because the stable Koszul complex is flat and P is Cohen-Macaulay) thus

$$H_I^i(M) = H_{r-i}(H_I^r(F_*)).$$

When $M = T$, we have a finitely generated free resolution whose terms F_i each have generators in a single degree d_i and are therefore Gorenstein:

$$H_I^r(F_i) = \text{Start}(d_i + 2r)F_i^\vee.$$

However, local cohomology is covariant, so the maps in the complex are not duals. We may revert to the case of a general finitely generated module for the answer.

LEMMA 4.7.1. (Local duality) *The $2r$th desuspension of the complex*

$$H_I^r(F_r) \longrightarrow H_I^r(F_{r-1}) \longrightarrow \cdots \longrightarrow H_I^r(F_0)$$

is obtained from

$$F_r \longrightarrow F_{r-1} \longrightarrow \cdots \longrightarrow F_0$$

by taking P-duals and then $\mathbb{F}_2$-duals, that is by applying

$$\text{Hom}_{\mathbb{F}_2}(\text{Hom}_P(\cdot, P), \mathbb{F}_2))$$

REMARK 4.7.2. Of course $\mathbb{F}_2$-duality is exact, so the essential step is the initial P-duality. The more compact formulation of the lemma is the statement

$$H_I^i(M) = \text{Ext}^{r-i}(M, P)^\vee(2r),$$

and it is this statement that is usually known as local duality, but for calculational purposes we use the form in the lemma.

Proof: This is an exercise in duality. If we choose bases and represent the maps by a matrix it is enough to consider a map $x : P(a) \longrightarrow P(b)$ given by multiplication with an element x of P (of degree $a - b$). The matrix of the new complex has this replaced by

$$(\cdot x)^* : \text{Start}(a + 2r)P^\vee \longrightarrow \text{Start}(b + 2r)P^\vee,$$

where the effect on $f \in \text{Start}(a + 2r)P^\vee = \text{Hom}(\Sigma^{a+2r}P, \mathbb{F}_2)$ is

$$(\cdot x)^*(f)(y) = f(xy).$$

Applying $\mathbb{F}_2$ duals we recover the original matrix. $\square$

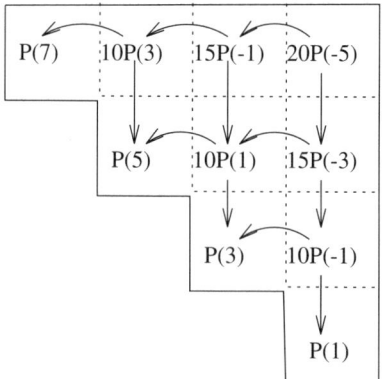

FIGURE 4.5. The double complex for the local cohomology of T_3 in rank 6.

From this we can deduce the local cohomology of T. As before, it is convenient to phrase the result in terms of the dual local cohomology modules, since these are Noetherian. It is worth alerting the reader to the fact that each of the statements in the following theorem represents quite exceptional behaviour, and left the authors incredulous on discovery. The second and third statements were suggested by calculations with the commutative algebra package Cocoa [9].

THEOREM 4.7.3. *The local cohomology of T_i is as follows*
  (1) *T_i only has local cohomology in degrees i and r.*
  (2) *For $i = 2, 3, \ldots, r - 1$*
$$H_I^r(T_i)^\vee = T_{r-i+2}(-r+4)$$
  and
$$H_I^r(T_r)^\vee = P(-r+4).$$
  (3) *The dual of $H_I^i(T_i)$ is only one dimensional, and has Hilbert series*
$$[H_I^i(T_i)^\vee] = t^{3i-2r-4}((1+x)^r - x^{r-i+1})/(1-x)$$
  where $x = t^2$ and t is of degree -1.

REMARK 4.7.4. It follows from Part 2 that if we take $T' = \bigoplus_3^{r-1} T_i$ then
$$H_I^r(T')^\vee = T'(-r+4)$$
so that T' is quasi-Gorenstein (the 'quasi' refers to the fact that T' is not of depth r). In fact the inclusion $T_2(-r+4) \longrightarrow P(-r+4)$ has a one dimensional quotient, so that the error in replacing T' by T in the above statement is only one dimensional. Furthermore, since the lower local cohomology is only one dimensional, T itself is Gorenstein in codimension $\leq r - 2$.

Proof: To calculate the local cohomology of T_i we consider the resolution $S = S(T_i)$ from the proof of 4.6.3. By 4.7.1 we need only reverse the direction of the arrows and change the suspensions to obtain a complex $S^\vee$ calculating the local cohomology. This immediately gives an algorithm for calculating it.

4.7. THE LOCAL COHOMOLOGY OF T.

LEMMA 4.7.5. *The extreme local cohomology modules of T_i are described as follows.*
(i) The bottom local cohomology $H^i_I(T_i)^\vee$ has a presentation

$$rP(2-i) \oplus rP(4-i) \oplus \cdots \oplus rP(2r-3i) \longrightarrow$$
$$P(4-i) \oplus P(6-i) \oplus \cdots \oplus P(2r-3i+4) \longrightarrow H^i_I(T_i)^\vee \longrightarrow 0;$$

in particular it is generated by elements in degrees $4-i, 6-i, \ldots, 2r-3i+4$.
(ii) The top local cohomology lies in an exact sequence

$$0 \longrightarrow H^r_I(T_i)^\vee \longrightarrow \binom{r}{i} P(4-i) \longrightarrow \binom{r}{i+1}[P(6-i) \oplus P(8-i)].$$

The map is described by

$$\xi_k(S) \longmapsto \sum_{t \notin S}(y_t \xi_{k+2}(S \cup \{t\}) + y_t^2 \xi_{k+4}(S \cup \{t\})).$$

To see that T_i only has local cohomology in degrees i and r, we we may use the same argument as in 4.6.3, since the dual of a Koszul complex is again a Koszul complex. The only difference is that we are truncating the Koszul complex at the other end: the Koszul complex only has homology in degree 0, so if we remove the lower modules (as in 4.6.3) it still only has homology at the bottom end. If we remove the upper end (as here) it retains its homology in degree 0, but also has homology at the upper end.

This completes the proof of Part 1. Since $T_r \cong P(-r-4)$, the statement about its local cohomology is clear, and we may suppose $i \leq r-1$. Part 2 is therefore an immediate consequence of the following.

LEMMA 4.7.6. *If $i = 2, 3, \ldots, r-1$ then there is an exact sequence*

$$0 \longrightarrow T_{r-i+2} \longrightarrow \binom{r}{r-i} P(i-r) \longrightarrow \binom{r}{r-(i+1)}[P(i-r+2) \oplus P(i-r+4)],$$

where the map is described by

$$\xi_k(S) \longmapsto \sum_{t \notin S} y_t \xi_{k+2}(S \cup \{t\}) + y_t^2 \xi_{k+4}(S \cup \{t\}).$$

Proof: If we choose as basis of $\binom{r}{r-i} P(i-r)$ the products $x_{i(1)} x_{i(2)} \cdots x_{i(r-i)}$ of $r-i$ distinct x's then the inclusion of T_{r-i+2} is standard. The exactness of the sequence now follows exactly as in 4.6.1. It is a zero-sequence since $Q_0(q_S) = Q_0(Q_1 Q_0 x_S) = 0$ and $Q_1(q_S) = Q_1(Q_1 Q_0 x_S) = 0$. It has no homology because $\ker(Q_1) \cap \ker(Q_0) = \operatorname{im}(Q_1 Q_0) + P$. Since $i \neq r$, the term P makes no contribution. □

To complete the proof of 4.7.3, it remains to establish the Hilbert series for $H^i_I(T_i)$ and hence its dimension. This is done in the following section. □

4.8. Hilbert series.

We shall be discussing Hilbert series of Noetherian modules over the polynomial ring P. We write
$$[M] = \sum_n t^n \dim(M_{-n}).$$
We have chosen t of degree -1, so this is a Laurent series in t. We let $x = t^2$ so that $[P] = 1/(1-t^2)^r = 1/(1-x)^r$. We shall only be discussing Hilbert series of Noetherian modules, so the Hilbert series is a rational function of t. Indeed, it is immediate from a resolution that the Hilbert series takes the form $[M] = p(t)/(1-x)^r$ for some polynomial $p(t)$. This section consists of entirely elementary manipulations with rational functions.

DEFINITION 4.8.1. For $0 \leq i \leq r$, we define truncations of the polynomial $(1-x)^r$ by
$$(1-x)^r_{[i]} = \binom{r}{i}(-x)^i + \binom{r}{i+1}(-x)^{i+1} + \cdots + \binom{r}{r}(-x)^r.$$

The identity $(-x)^r(1-(1/x))^r = (1-x)^r$ gives a useful duality property for the truncations.

LEMMA 4.8.2. The truncated binomials have the following duality property
$$(-x)^r(1-(1/x))^r_{[i]} + (1-x)^r_{[r-i+1]} = (1-x)^r. \quad \square$$

LEMMA 4.8.3. The Hilbert series of T_i is
$$[T_i] = (-t)^{-i+4}((1-x)^r_{[i]} - x^{1-i}(1-x^2)^r_{[i]})/(1-x)^{r+1}.$$

Proof: Directly from the resolution in 4.6.3, we calculate
$$\begin{aligned}
(1-x)^r[T_i] &= \binom{r}{i}t^{i+4} - \binom{r}{i+1}(t^{i+6}+t^{i+8}) + \binom{r}{i+2}(t^{i+8}+t^{i+10}+t^{i+12}) - \cdots \\
&= (-t)^{-i+4}[\binom{r}{i}(-x)^i + \binom{r}{i+1}(-x)^{i+1}(1+x) \\
&\quad + \binom{r}{i+2}(-x)^{i+2}(1+x+x^2) + \cdots] \\
&= (-t)^{-i+4}[\binom{r}{i}(-x)^i(1-x) + \binom{r}{i+1}(-x)^{i+1}(1-x^2) \\
&\quad + \binom{r}{i+2}(-x)^{i+2}(1-x^3) + \cdots]/(1-x) \\
&= (-t)^{-i+4}[(\binom{r}{i}(-x)^i + \binom{r}{i+1}(-x)^{i+1} + \cdots) \\
&\quad - (1/x^{i-1})(\binom{r}{i}(-x^2)^i + \binom{r}{i+1}(-x^2)^{i+1} + \cdots)]/(1-x) \\
&= (-t)^{-i+4}[(1-x)^r_{[i]} - x^{1-i}(1-x^2)^r_{[i]}]/(1-x)
\end{aligned}$$

as required. $\quad \square$

We are now ready to deduce the Hilbert series of various dual local cohomology modules. It is worth reminding readers that the ideal behaviour enjoyed by Gorenstein modules M is that one should have $[H^r_I(M)^\vee] = (-1)^r t^j [M](1/t)$ for some j.

LEMMA 4.8.4. For $i = 2, 3, \ldots, r-1$ we have
$$[H^r_I(T_i)^\vee] = (-1)^r[T_i](1/t) - (-1)^{r-i}[H^i_I(T_i)^\vee].$$

Proof: Take the resolution F_* of T_i from 4.6.3. We have $[T_i] = \chi([F_*])$. By local duality 4.7.1, the cohomology of the $2r$-th desuspension of the dual of F_* is the local cohomology. Finally, by 4.7.3, since the local cohomology is only in degrees i and r we find

$$[H_I^r(T_i)^\vee] + (-1)^{r-i}[H_I^i(T_i)^\vee] = \chi([\text{Hom}(F_*, P)])$$
$$- (-1)^r \chi([F_*])(1/t) = (-1)^r[M](1/t)$$

as required. □

From the calculation in 4.7.3 we deduce the Hilbert series of $H_I^i(T_i)^\vee$.

COROLLARY 4.8.5. *For $i = 2, 3, \ldots, r-1$ we have*

$$[H_I^i(T_i)^\vee] = t^{3i-2r-4}((1+x)^r - x^{r-i+1})/(1-x).$$

Proof: Combining 4.7.3 Part 2 with 4.8.4, we find

$$(-1)^{r-i}[H_I^i(T_i)^\vee] = (-1)^r[T_i](1/t) - t^{r-4}[T_{r-i+2}].$$

Let us record first that by 4.8.3

$$t^{r-4}[T_{r-i+2}] = (-1)^{r-i}t^{i-2}[(1-x)^r_{[r-i+2]} - x^{i-r-1}(1-x^2)^r_{[r-i+2]}]/(1-x)^{r+1}$$
$$= (-1)^{r-i}t^{i-2}[(1-x)^r_{[r-i+1]} - x^{i-r-1}(1-x^2)^r_{[r-i+1]}]/(1-x)^{r+1}.$$

Now we use 4.8.3 and 4.8.2 to deduce

$$(-1)^r[T_i](\tfrac{1}{t}) = (-1)^r(-t)^{i-4}[(1-\tfrac{1}{x})^r_{[i]} - x^{i-1}(1-\tfrac{1}{x^2})^r_{[i]}]/(1-\tfrac{1}{x})^{r+1}$$
$$= -(-t)^{i-4}[x^{r+1}(1-\tfrac{1}{x})^r_{[i]} - x^{r+i}(1-\tfrac{1}{x^2})^r_{[i]}]/(1-x)^{r+1}$$
$$= -(-t)^{i-4}[x(-1)^r((1-x)^r - (1-x)^r_{[r-i+2]})$$
$$- x^{i-r}(-1)^r((1-x^2)^r - (1-x^2)^r_{[r-i+2]})]/(1-x)^{r+1}$$
$$= (-1)^{r-i}t^{i-4}x[-((1-x)^r - x^{i-r-1}(1-x^2)^r)$$
$$+ ((1-x)^r_{[r-i+1]} - x^{i-r-1}(1-x^2)^r_{[r-i+1]})]/(1-x)^{r+1}).$$

Subtracting $t^{r-4}[T_{r-i+2}]$ we obtain the desired result. □

4.9. The quotient P/T_2.

When V is of rank 1, the module T_2 is zero, so we suppose $r \geq 2$ for the rest of the section. When V is of rank 2, we see $T \cong \Sigma^{-6}P$ (and quickly check this isomorphism is untwisted).

The confusing thing about T_2 is that its generators are not monomials, so we let

$$\hat{T} = (y_i y_j \mid i < j)$$

and define U and A by the exact sequences

$$0 \longrightarrow T \longrightarrow \hat{T} \longrightarrow U \longrightarrow 0$$

and

$$0 \longrightarrow \hat{T} \longrightarrow P \longrightarrow A \longrightarrow 0.$$

First note that A is the 'axis quotient' of P (of dimension 1 in degree 0 and of dimension r for each negative even degree).

Now we turn to the module U. This has an $\mathbb{F}_2$ basis consisting of elements y_S^n of degree $-2n$, where S is a subset of the possible indices $\{1, 2, \ldots, r\}$ with at least two elements, and $n \geq |S|$:
$$U = (y_S^n \mid n \geq |S| \geq 2).$$
The action of P is given by
$$y_i y_S^n = y_{S \cup \{i\}}^{n+1}.$$
This is very reminiscent of the description of Q', the formal difference being that we are now working mod $(2, v)$. The principal difference in detail is that only subsets with 2 or more elements give rise to basis elements.

Now we have a $GL(V)$-invariant filtration
$$U = U_2 \supset U_3 \supset \cdots \supset U_r \supset U_{r+1} = 0$$
where
$$U_i = (y_S^n \mid |S| \geq i).$$

LEMMA 4.9.1. *The subquotient U_i/U_{i+1} is non-zero in negative even dimensions $\leq -2i$, and is of dimension $\binom{r}{i}$ in each such degree.* □

We leave the reader the straightforward exercise of using the contents of this section to give an alternative proof that T_2 only has local cohomology in degrees 2 and r, and that $H_I^r(T_2) = P(-r+4)$ provided $r \geq 3$. One ingredient is that the element y^* is U-regular. The disadvantage of this method is that it only gives $H_I^2(T_2)$ up to extension.

4.10. The local cohomology of R.

We reassemble the local cohomology of R from the short exact sequence
$$0 \longrightarrow T \longrightarrow R \longrightarrow Q \longrightarrow 0.$$
First, Q is of dimension 1, and so has local cohomology only in degrees 0 and 1. Next $T = T_2 \oplus T_3 \oplus \cdots \oplus T_r$, and T_i is of depth i and dimension r; by 4.7.3 T_i only has local cohomology in degrees i and r. Accordingly the only possible non-trivial connecting homomorphism is
$$\delta: H_I^1(Q) \longrightarrow H_I^2(T_2).$$
In fact we shall see in 4.11.5 that this is surjective with kernel equal to $2 \cdot H_I^1(Q)$.

To avoid special cases, in the general discussion, we make explicit the examples $r \leq 3$.

EXAMPLE 4.10.1. (i) If $r = 2$ then $H_I^0(R) = ku^* \cdot \rho$ and there is an exact sequence
$$0 \longrightarrow H_I^1(R) \longrightarrow H_I^1(Q) \overset{\delta}{\longrightarrow} \mathrm{Start}(-2)P^\vee \longrightarrow H_I^2(R) \longrightarrow 0.$$
The map δ is determined by its effect in degree -2, since the codomain is dual of a free module on a single generator of degree 2. In this case $H_I^1(Q)$ is one dimensional in degree -2, so it suffices to show it is non-trivial. We may argue using topology. Indeed, the local cohomology spectral sequence collapses, so we need to ensure $H_I^2(R)$ is zero below degree 2, since $ku_*(BV)$ is connective. The map δ is calculated on $x \in H_I^1(Q)_{-4+2n}$ by taking $\delta(x) \in P_{-2+2n}^\vee$ to be the map

$P_{-2n+2} \longrightarrow \mathbb{F}_2$ given by acting on x. A short calculation allows us to deduce δ is an isomorphism in degrees 0 and -2, and in general
$$H_I^1(R) = 2 \cdot H_I^1(Q)$$
and
$$H_I^2(R) = \text{Start}(4)(y_1^2 y_2 + y_1 y_2^2)^{\vee} \cong P^{\vee}(4).$$
(ii) If $r = 3$ then $H_I^0(R) = ku^* \cdot \rho$ and there is an exact sequences
$$0 \longrightarrow H_I^1(R) \longrightarrow H_I^1(Q) \xrightarrow{\delta} H_I^2(T) \longrightarrow H_I^2(R) \longrightarrow 0$$
and
$$H_I^3(R) = H_I^3(T_2) \oplus H_I^3(T_3) = P^{\vee}(6) \oplus P^{\vee}(-1).$$
In fact we shall see that δ is surjective, and hence
$$H_I^i(R) = \begin{cases} ku^* \cdot \rho & \text{if } i = 0 \\ 2 \cdot H_I^1(Q) & \text{if } i = 1 \\ 0 & \text{if } i = 2 \\ P^{\vee}(6) \oplus P^{\vee}(-1) & \text{if } i = 3 \end{cases}$$

The general result is as follows.

PROPOSITION 4.10.2. *The local cohomology of R is*
$$H_I^i(R) = \begin{cases} ku^* \cdot \rho & \text{if } i = 0 \\ 2 \cdot H_I^1(Q) & \text{if } i = 1 \\ 0 & \text{if } i = 2 \\ H_I^i(T_i) & \text{if } 3 \leq i \leq r-1 \\ H_I^r(T) & \text{if } i = r \end{cases}$$

REMARK 4.10.3. By 4.7.3 we have a presentation of $H_I^i(T_i)$, we know its Hilbert series and that it is one dimensional. We also know
$$H_I^r(T)^{\vee} = (P \oplus T_3 \oplus T_4 \oplus \cdots \oplus T_r)(-r+4).$$

Proof: This is immediate except for degrees 1 and 2. For these, it suffices to identify the connecting homomorphism.

LEMMA 4.10.4. *The map $\delta : H_I^1(Q) \longrightarrow H_I^2(T_2)$ is surjective.*

We will prove a more general statement in 4.11.5 below.

It then immediately follows that $H_I^2(R) = 0$. Furthermore $H_I^1(R) = 2 \cdot H_I^1(Q)$ since certainly $2 \cdot H_I^1(Q)$ lies in the kernel of δ. On the other hand $H_I^1(Q)/2$ and $H_I^2(T_2)$ are both graded $\mathbb{F}_2$ vector spaces of the same finite dimension in each degree: their duals both have Hilbert series $t^{2-2r}((1+x)^r - x^{r-1})/(1-x)$ by 4.5.1 and 4.7.3. $\square$

4.11. The ku-homology of BV.

Before treating the general case, we make the cases of rank 1, 2 and 3 explicit. This should help the reader to get a picture of the spectral sequence, as will be necessary follow the general argument.

EXAMPLE 4.11.1. If V is of rank 1 the local cohomology spectral sequence obviously collapses. Hence $ku_{ev}(BV) = H_I^0(R) = ku^* \cdot \rho$, and $ku_{od}(BV) = \Sigma^{-1} H_I^1(R)$. This is cyclic of order 2^n in dimension $2n - 1$.

EXAMPLE 4.11.2. Quite generally, we know $H^0_I(R) = ku^* \cdot \rho$ survives. Hence the spectral sequence also collapses if V is of rank 2. Hence

$$ku_{ev}(BV) = H^0_I(R) \oplus \Sigma^{-2} H^2_I(R) = ku^* \cdot \rho \oplus \text{Start}(2) P^\vee$$

and

$$ku_{od}(BV) = \Sigma^{-1} H^1_I(R) = 2 \cdot H^1_I(Q),$$

which is

$$\binom{2}{1} \mathbb{Z}/2^n \oplus \binom{2}{2} \mathbb{Z}/2^{n-1}$$

in degree $2n - 1$.

Because of the grading in the local cohomology spectral sequence, if $H^i_I(R)$ is non-zero below degree i there must be a differential. Accordingly the spectral sequence does not collapse for $r \geq 3$.

EXAMPLE 4.11.3. In the rank 3 case, the differential can be inferred from this information. Indeed, since $H^1_I(R)$ is in even degrees, the differential is

$$d^2 : H^1_I(R) \longrightarrow \Sigma^{-1}(H^3_I(R)_{od}).$$

Since the codomain $\Sigma^{-1}(H^3_I(R)_{od}) = \Sigma^{-1} H^3_I(T_3) = \Sigma^{-2} P^\vee$ is the dual of a free module, the differential is determined by its restriction to the bottom non-zero degree (namely $H^1_I(R)_{-2}$) where it must be an isomorphism since $ku_*(BV)$ is connective. A short calculation allows us to deduce

$$\text{cok}(d^2) = \text{Start}(10)(y_1^2 y_2 + y_1 y_2^2, y_1^2 y_3 + y_1 y_3^2, y_2^2 y_3 + y_2 y_3^2)^\vee = \text{Start}(10) T_2^\vee$$

and

$$\ker(d^2) = 4 \cdot H^1_I(Q)$$

Accordingly,

$$\begin{aligned} ku_{ev}(BV) &= H^0_I(R) \oplus \Sigma^{-2} H^2_I(R) \oplus \Sigma^{-3}(H^3_I(R)_{od}/d^2) \\ &= ku^* \cdot \rho \oplus \Sigma^{-3}(H^3_I(T_3)_{od}/d^2) \\ &= ku^* \cdot \rho \oplus \text{Start}(8) T_2^\vee \end{aligned}$$

For $ku_{od}(BV)$ we have an exact sequence

$$0 \longrightarrow \Sigma^{-3}(H^3_I(R)_{ev}) \longrightarrow ku_{od}(BV) \longrightarrow \Sigma^{-1} \ker(d^2) \longrightarrow 0,$$

and since $H^3_I(R)_{ev} = H^3_I(T_2) \cong P^\vee(6)$, a split exact sequence

$$0 \longrightarrow P^\vee(3) \longrightarrow ku_{od}(BV) \longrightarrow \text{Start}(1)(4 \cdot H^1_I(Q)) \longrightarrow 0.$$

From these descriptions it is easy to identify the abelian group in each degree, and the action of $GL(V)$, at least up to extension in the final case.

We are now ready to describe the behaviour of the spectral sequence in general.

THEOREM 4.11.4. The spectral sequence has E^∞-term on the columns $s = 0, -1$ and $-r$, and this gives an extension

$$0 \longrightarrow \text{Start}(2) T^\vee \longrightarrow \widetilde{ku}_*(BV) \longrightarrow \text{Start}(1)(2^{r-1} H^1_I(Q)) \longrightarrow 0$$

of $GL(V)$-modules, and the extension is additively split.

Proof: We prove the theorem by showing that the only differentials are those originating on the $s = -1$ column and all but the last of these are epimorphisms. We may include the connecting homomorphism $\delta : H_I^1 Q \longrightarrow H_I^2 T_2$ in the discussion by treating it as d^1.

The most obvious fact is that the codomain of each differential is annihilated by 2, and so the kernel of each differential includes all multiples of 2. We show by induction that this is precisely the kernel, and that there are no other differentials. Suppose then that this has been proved for differentials $d_1, \ldots, d_{i-2}$ (the assumption is vacuous for $i = 2$).

We choose to index our induction by the degree of local cohomology involved as the codomain. Thus at the ith stage we have

$$d^{i-1} : 2^{i-2} H_I^1(Q) \longrightarrow H_I^i(T_i)(2-i),$$

and this factors through the $\mathbb{F}_2$-vector space $2^{i-2} H_I^1(Q)/2^{i-1} H_I^1(Q)$. It is natural to replace this map between Artinian modules by the dual map

$$d'_{i-1} : H_I^i(T_i)^{\vee}(i-2) \longrightarrow (2^{i-2} H_I^1(Q)/2^{i-1} H_I^1(Q))^{\vee}$$

between Noetherian modules.

We may now treat the cases $i = 2, \ldots, r-1$ together, but the case $i = r$ is slightly different and we return to it at the end.

PROPOSITION 4.11.5. *For $i = 2, \ldots, r-1$ the dual differential*

$$d'_{i-1} : H_I^i(T_i)^{\vee}(i-2) \longrightarrow (2^{i-2} H_I^1(Q)/2^{i-1} H_I^1(Q))^{\vee}$$

is an isomorphism.

Proof: By 4.7.3 and 4.5.1 these two modules have the same Hilbert series if $i \leq r-1$ so that if the map is an epimorphism it is an isomorphism.

By 4.5.2, it therefore suffices to show that d'_i is an epimorphism in positive degrees. The idea here is that the subgroup of $H_I^1(Q)$ in degree $t \leq 0$ must die in the spectral sequence, since it contributes to $ku_{t-1}(BV)$. The only differentials that affect the (-1)-column are the differentials d^i and the way to make the $E_{-1,t}^{i+1}$ term as small as possible is to ensure d^i is surjective. By counting we see that this gives exactly enough to kill $H_I^1(Q)_t$ with nothing left over: there are therefore no other differentials involving these groups, and the differentials d^i are all surjective.

LEMMA 4.11.6. *The order of $H_I^1(Q)$ in any degree $-2j \leq 0$ is exactly equal to the order of*

$$H_I^2(T_2)_{-2j+1} \oplus H_I^3(T_3)_{-2j+2} \oplus \cdots \oplus H_I^r(T_r)_{-2j+r-1},$$

and $H_I^r(T_i)_{-2j+r-1} = 0$ for $i \leq r-1$. Accordingly, all differentials d^i on $H_I^1(Q)_{-2j}$ are surjective.

This completes the proof of 4.11.5. $\square$

Finally, we must understand

$$d'_{r-1} : P(2) = P(4-r)(r-2) = H_I^r(T_r)^{\vee}(r-2) \longrightarrow (2^{r-2} H_I^1(Q)/2^{r-1} H_I^1(Q))^{\vee}.$$

Evidently the map is determined by the image of the generator in degree 2, and it is easily verified that the codomain is one dimensional in that degree. Since $ku_*(BV)$

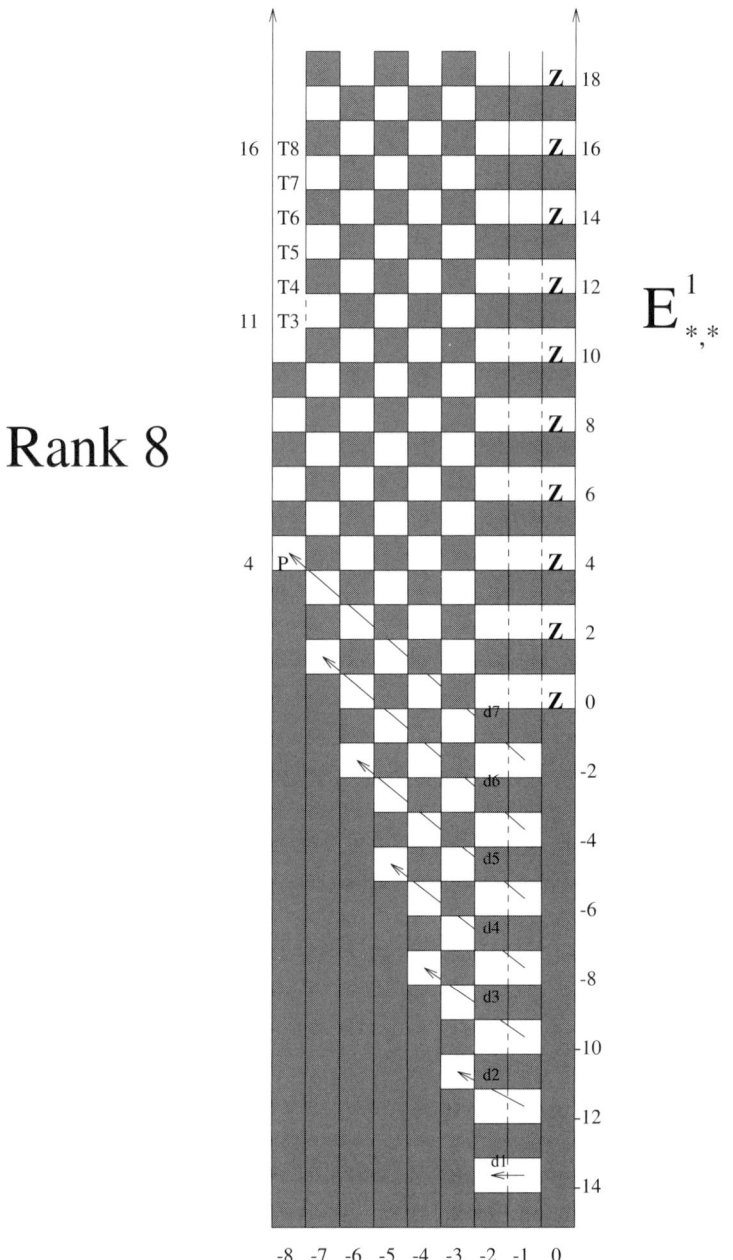

FIGURE 4.6. **The local cohomology spectral sequence for $ku_*(BV)$ with V of rank 8.** The (-1)st and (-2)nd columns start in degrees $-2r+2$. After this, the start increases in steps of 3. The bottom nonzero entry of the $(-r)$th column is in degree $r-4$ (coming from $H_I^r(T_r) = P^\vee(r-4)$). The subsequent contributions are $\text{Start}(r+3)T_3^\vee$, $\text{Start}(r+4)T_4^\vee$, ..., $\text{Start}(2r)T_r^\vee = P^\vee(2r)$. In the picture, the differential d^i connects the lowest nonzero groups in their respective columns on the E^i-page.

4.12. DUALITY FOR THE COHOMOLOGY OF ELEMENTARY ABELIAN GROUPS.

is connective, the d'_{r-1} is surjective in degree 2, and hence by 4.5.2 it is surjective in all degrees.

LEMMA 4.11.7. *The kernel of d'_{r-1} is precisely $T_2(2) \subseteq P(2)$.*

Proof: Since the two monomials in a generator $y_i^2 y_j + y_i y_j^2$ of T_2 involve the same set of generators $\{y_i, y_j\}$ and are of the same degree they map to the same place in the module $(2^{r-2} H_I^1(Q)/2^{r-1} H_I^1(Q))^\vee$. Thus $\ker(d'_{r-1})$ contains T_2. On the other hand we see from 4.8.3 and 4.5.1 that P/T_2 and $(2^{r-2} H_I^1(Q)/2^{r-1} H_I^1(Q))^\vee$ have the same Hilbert series. $\square$

Finally, we have not established that the components of d^{r-1} involving $H_I^r(T_i)$ with $i \leq r-1$ are non-zero. However, since d'_{r-1} is surjective, the E^∞ term is determined up to isomorphism. $\square$

4.12. Duality for the cohomology of elementary abelian groups.

In this section we implement the discussion from Section 3.3 for elementary abelian groups. It is generally considered that the homology of a group is more complicated than its cohomology since it involves various forms of higher torsion. However, one of the lessons of the local cohomology approach is that, after duality, the two contain very similar information. In fact the local cohomology spectral sequence is a manifestation of a remarkable duality property of the ring $ku^*(BG)$. For ordinary mod p cohomology the corresponding duality implies, for example, that a Cohen-Macaulay cohomology ring is automatically Gorenstein. Since ku^* is more complicated than $H\mathbb{F}_p^*$, the statement is more complicated for connective K theory, but the phenomenon is nonetheless very striking. This is reflected again in Tate cohomology. As with ordinary cohomology, the advantages of Tate cohomology are most striking in the rank 1 case.

More precisely, we view the local cohomology theorem
$$H_I^*(ku^*(BG)) \Rightarrow ku_*(BG)$$
as the statement
$$R\Gamma_I(ku^*(BG)) = ku_*(BG),$$
and the universal coefficient spectral sequence
$$\operatorname{Ext}_{ku_*}^{*,*}(ku_*(BG), ku_*) \Rightarrow ku^*(BG),$$
as the statement
$$RD_{ku_*}(ku_*(BG)) = ku^*(BG)$$
where $D_{ku_*}(\cdot) = \operatorname{Hom}_{ku_*}(\cdot, ku_*)$ denotes duality, and R denotes the right derived functor in some derived category. Combining these two we obtain a statement of the form
$$(RD_{ku_*}) \circ (R\Gamma_I)(ku^*(BG)) = ku^*(BG).$$
This states that the ring $ku^*(BG)$ is "homotopically Gorenstein" in some sense. We give this heuristic discussion substance by showing what this means in practice for elementary abelian groups.

Remarkably, it is a profound reflection of the obvious formal similarity between the sequences
$$0 \longrightarrow T \longrightarrow ku^*(BV) \longrightarrow Q \longrightarrow 0$$

and
$$0 \longrightarrow \mathrm{Start}(2)T^\vee \longrightarrow \widetilde{ku}_*(BV) \longrightarrow \mathrm{Start}(1)(2^{r-1}H^1_I(Q)) \longrightarrow 0.$$

It turns out these sequences correspond closely to the homological properties of $ku^*(BV)$ as a module over ku^*. It is convenient to remove the summand ku^* from both homology and cohomology: since $\mathrm{Hom}_{ku_*}(ku_*, ku_*) = ku_*$, this fits into the universal coefficient spectral sequence pattern as described in Section 3.2, with
$$S = \mathrm{Start}(2)T^\vee \text{ and } P = \mathrm{Start}(1)(2^{r-1}H^1_I(Q)).$$

As remarked in Section 3.2 the spectral sequence collapses to the short exact sequence
$$0 \longrightarrow \mathrm{Ext}^2_{ku_*}(\Sigma^2 S, ku_*) \longrightarrow \widetilde{ku}^*(BV) \longrightarrow \mathrm{Ext}^1_{ku_*}(\Sigma\widetilde{P}, ku_*) \longrightarrow 0.$$

The main result of this section shows the resemblance is no coincidence. It is convenient have notation for the reduced quotient $\widetilde{Q} = Q/ku_*$, where ku_* is the submodule generated by 1.

PROPOSITION 4.12.1. *The short exact sequence for $\widetilde{ku}_*(BV)$ from the local cohomology spectral sequence is dual to the short exact sequence for $\widetilde{ku}^*(BV)$ from the universal coefficient theorem. More precisely*
$$\mathrm{Ext}^i_{ku_*}(\mathrm{Start}(4)T^\vee, ku_*) = \begin{cases} T & \text{if } i = 2 \\ 0 & \text{otherwise} \end{cases}$$

and
$$\mathrm{Ext}^i_{ku_*}(\mathrm{Start}(2)(2^{r-1}H^1_I(Q)), ku_*) = \begin{cases} \widetilde{Q} & \text{if } i = 1 \\ 0 & \text{otherwise} \end{cases}$$

Proof: We essentially repeat the proof of 3.2.1 with the relevant particular values of S and P, and then continue far enough to calculate the non-vanishing Ext groups.

It is natural to use the injective resolution
$$0 \longrightarrow ku^* \longrightarrow ku^*[1/p, 1/v] \longrightarrow$$
$$ku^*[1/p]/v^\infty \oplus ku^*[1/v]/p^\infty \longrightarrow ku^*/(p^\infty, v^\infty) \longrightarrow 0$$
of ku^*.

The statement about $T^\vee$ then follows directly. Since $T^\vee$ is 2 and v torsion, it follows that
$$\begin{aligned}
\mathrm{Ext}^i_{ku_*}(\mathrm{Start}(4)T^\vee, ku_*) &= \mathrm{Ext}^2_{ku_*}(\mathrm{Start}(4)T^\vee, ku_*) \\
&= \mathrm{Hom}_{ku_*}(\mathrm{Start}(4)T^\vee, ku^*/p^\infty, v^\infty) \\
&= \mathrm{Hom}_{\mathbb{F}_2}(\mathrm{Start}(4)T^\vee, \Sigma^{-2}\mathbb{F}_2) \\
&= \mathrm{Start}(-6)T = T.
\end{aligned}$$

Now consider $M = \mathrm{Start}(2)M = 2^{r-1}H^1_I(Q)$. First note that it is 2-power torsion and hence its Ext groups are the cohomology of the sequence
$$0 \xrightarrow{d^0} \mathrm{Hom}_{ku_*}(M, ku^*[1/v]/p^\infty) \xrightarrow{d^1} \mathrm{Hom}_{ku_*}(M, ku^*/p^\infty, v^\infty).$$

For the term in cohomological degree 1, we note
$$\begin{aligned}
\mathrm{Hom}_{ku_*}(M, ku^*[1/v]/p^\infty) &= \mathrm{Hom}_{ku_*}(M[1/v], ku^*[1/v]/p^\infty) \\
&= \mathrm{Hom}_{\mathbb{Z}}(M[1/v]_0, \mathbb{Z}/p^\infty).
\end{aligned}$$

However $M[1/v] = (K^*(BV)/\rho)/p^\infty$, so the resulting group of homomorphisms is $(R(V)/(\rho))^\wedge_p$ in each even degree.

For the term in cohomological degree 2, we note that M starts in degree 2, and that for $k \geq 0$ multiplication by v includes M_{2k} in M_{2k+2} as $2M_{2k+2}$. Hence
$$\mathrm{Hom}^{2k}_{ku_*}(M, ku^*/p^\infty, v^\infty) = \mathrm{Hom}_\mathbb{Z}(M_{2k}, \mathbb{Z}/p^\infty).$$
Next note that M_{2k} contributes to $ku_{2k-1}(BV)$, and maps monomorphically under inverting v to
$$K_{2k-1}(BV) = H^1_I(R(V)) = (R(V)/\rho)/p^\infty.$$
Hence the differential d^1 is surjective. It follows that $\mathrm{Ext}^2_{ku_*}(M, ku_*) = 0$, and that
$$\mathrm{Ext}^{1,2k}_{ku_*}(M, ku_*) = \mathrm{Hom}_\mathbb{Z}([(R(G)/\rho)/p^\infty]/M_{2k}, \mathbb{Z}/p^\infty).$$
As an abelian group this is certainly right, but we should also consider the action of v. We have shown that
$$\mathrm{Ext}^{1,2k+2}_{ku_*}(M, ku_*) \xrightarrow{v} \mathrm{Ext}^{1,2k}_{ku_*}(M, ku_*)$$
corresponds to
$$v : M_{2k+2} \xleftarrow{v} M_{2k}.$$
The former is therefore an isomorphism for $k \leq -1$, and a monomophism with cokernel $M_{2k+2}/2$ for $k \geq 0$. This precisely corresponds to the quotient
$$\widetilde{Q}^{2k+2} \xrightarrow{v} \widetilde{Q}^{2k}$$
of
$$ku^{2k+2}(BV) \xrightarrow{v} ku^{2k}(BV). \quad \square$$

4.13. Tate cohomology of elementary abelian groups.

In this section we combine our calculations of the ku-homology and cohomology of BV with the completely known norm sequence for periodic K theory to deduce $t(ku)^*_V$, at least up to extension.

PROPOSITION 4.13.1. *If V is an elementary abelian 2-group there is an extension of $ku^*(BV)$-modules*
$$0 \longrightarrow T \oplus \mathrm{Start}(2)T^\vee \longrightarrow t(ku)^*_V \longrightarrow N \longrightarrow 0$$
where N has no v-torsion and is additively $(\mathbb{Z}_2^\wedge)^{|V|-1}$ in each even degree.

Proof: There is certainly an exact sequence of the form
$$0 \longrightarrow \Gamma_v t(ku)^*_V \longrightarrow t(ku)^*_V \longrightarrow N \longrightarrow 0,$$
and we will show this is as stated in the proposition.

The norm sequence gives
$$0 \longrightarrow ku^*(BV)/(\rho) \longrightarrow t(ku)^*_V \longrightarrow \Sigma \widetilde{ku}_*(BV) \longrightarrow 0.$$

We compare this with the norm sequence for periodic K, noting that by 1.1.1, the latter is obtained by inverting v. By, 4.2.7, the v-power torsion in $ku^*(BV)/(\rho)$ is T, and by 4.11.4, the v-power torsion in $\widetilde{ku}_*(BV)$ is $S = \mathrm{Start}(2)T^\vee$. Since the v-power torsion functor is not right exact, we need a further argument to check all of S comes from v-power torsion in Tate cohomology. A comparison of the norm sequences for ku and K gives a six term exact sequence by the Snake Lemma. In this exact sequence, the group S (the kernel in homology) maps to the cokernel of $ku^*(BV)/(\rho) \longrightarrow K^*(BV)/(\rho)$. However, the latter is entirely in negative degrees since the cofibre of $ku \longrightarrow K$ is in negative degrees. Since S is in positive degrees,

the map in the snake sequence is zero, and the v-power torsion is an extension of T by S; since T is in degrees ≤ -6 and S is in degrees ≥ 2 we find
$$\Gamma_v t(ku)_V^* = T \oplus \mathrm{Start}(2)T^\vee,$$
as claimed.

Recall that by [30, 19.6] $t(K)_V^*$ is a $\mathbb{Q}_2$-vector space of dimension $|V| - 1$ in each degree. Its subgroup $N = t(ku)_V^*/\Gamma_v t(ku)_V^*$ therefore has no $\mathbb{Z}$-torsion. Since we have an exact sequence
$$0 \longrightarrow \widetilde{Q} \longrightarrow N \longrightarrow \widetilde{P} \longrightarrow 0,$$
and in each even degree $\widetilde{Q}$ is $(\mathbb{Z}_2^\wedge)^{|V|-1}$ and $\widetilde{P}$ is finite, the asserted structure of N follows. □

The proof of 4.13.1 gives further information about the product structure on $t(ku)_V^*$, and its structure as a module over $ku^*(BV)$. To start with, $N = t(ku)_V^*/\Gamma_v t(ku)_V^*$ is a subring of $t(K)_V^*$, and with the Adams spectral sequence we can identify it explicitly.

PROPOSITION 4.13.2. *(i) The image of $ku^*(BV)$ in $K^*(BV) = R(V)_J^\wedge[v, v^{-1}]$ is the Rees ring of $R(V)_J^\wedge$ for the ideal J. This is the subring generated by $\mathcal{E} = vJ$ and v.*
(ii) The image of
$$t(ku)_V^* \longrightarrow t(K)_V^* = R(V)_J^\wedge/(\rho)[1/2][v, v^{-1}]$$
is the subring generated by $\mathcal{E} = vJ$ and v together with $v^{r+n}\mathcal{E}^i/2^{n+1}$ for $n \geq 0$ and $1 \leq i \leq r-1$ and $v^{2r+n}\mathcal{E}^r/2^{n+r+1}$ for $n \geq 0$. In particular it is not finitely generated for $r \geq 2$.

REMARK 4.13.3. (i) The statement covers the case $r = 1$ when $t(ku)_V^* = \mathbb{Z}_2^\wedge[y, y^{-1}]$, since in this case $y^{-1} = v^2y/4$ is of filtration 0. This is exceptional in being a finitely generated algebra over $\mathbb{Z}_2^\wedge[y]$.

One may similarly make other cases explicit. For example, when $r = 2$,
$$N = \mathbb{Z}_2^\wedge[y_1, y_2, v^{n+1}y_1/2^{n+1}, v^{n+1}y_2/2^{n+1}, v^{n+2}y_1y_2/2^{n+2} \mid vy_i^2 = 2y_i, n \geq 0].$$

(ii) The reader is encouraged to identify these rings in the display of Q as in 4.4.2 and 4.3. This gives a picture of Part (i), since the image is Q. For Part (ii), the image of $t(ku)_V^*$ is the subring of $Q[1/y^*]$ consisting of $Q/(\rho)$ together with all elements of filtration $\geq r$. However, it is important to note the rather exceptional behaviour of $\mathcal{E}^r$. Indeed, since $\rho = 0$ the expression given before 4.4.3 gives
$$v^r\mathcal{E}^r \subseteq \sum_{i=1}^{r-1} v^{2r-i}\mathcal{E}^i/2^{r-i+1}.$$

Proof: Part (i) is immediate from the fact that $ku^*(BV)$ is generated as a ring by v and the Euler classes $y_1, \ldots, y_r$.

For Part (ii) we may include v and the Euler classes $y_1, \ldots, y_r$. Evidently it suffices to add classes that map to a generating set in $\Sigma\widetilde{P} = 2^{r-1}H_I^1(Q)$. □

We know various things about the action of N on $\Gamma_v t(ku)_V^*$. Firstly, it factors through $N/(2, v)N$. Secondly, it is consistent with the action of $\mathbb{F}_2[y_1, \ldots, y_r] = Q/(2, v)Q$ on T from $ku^*(BV)$. The action of the positive degree elements mapping

4.13. TATE COHOMOLOGY OF ELEMENTARY ABELIAN GROUPS.

onto the elements of P of order 2 are rather mysterious. Note that N preserves S, since the negative part of N is generated by degree -2 classes. Accordingly, the action of N on $S = \text{Start}(2)T^\vee$ follows by Tate duality.

Next the product of two elements of T is known, since it comes from cohomology, where the product comes from that in the polynomial ring $\mathbb{F}_2[x_1, \ldots, x_r]$, and the identification of S with $\text{Start}(2)T^\vee$ shows how T acts on S. Finally, in all cases with $T \neq 0$ (i.e., for $r \geq 2$), T has a regular sequence of length 2. Hence the arguments of Benson-Carlson [**7**, 2.1,3.1] apply precisely as in group cohomology (despite the unusual-looking dimension shifts) to give the triviality of the remaining products.

PROPOSITION 4.13.4. *If $s, s' \in S$ and $t \in T$ then $ss' = 0$ and if $|st| < 0$ then $st = 0$.* □

APPENDIX A

Conventions.

A.1. General conventions.

Unreduced: All homology and cohomology is unreduced unless the contrary is indicated by a tilde.

Degree and codegree: All homological (lower) indices are called degrees, and all cohomological (upper) indices are called codegrees. These are related by $M_i = M^{-i}$.

Mod p cohomology: Unless otherwise indicated, coefficients of ordinary cohomology are in $\mathbb{F}_p$ for some prime p.

Suspension: We have indicated suspension in three different ways, according to what is convenient. The algebraic and topological suspensions are related by
$$\Sigma^n M = M(n) = \Sigma_{-n} M.$$
We also write $\mathrm{Start}(n)M$ for the suspension in which the lowest non-zero entry is in degree n. Use of the notation implies that M is non-zero and bounded below.

Representations: (i) If H is the normal subgroup generated by s in G, with cyclic quotient G/H, then $\hat{s}$ denotes a one dimensional representation of G with kernel H.

(ii) The trivial one dimensional complex representation is denoted ϵ.

A.2. Adams spectral sequence conventions.

As usual our Adams spectral sequences
$$E_2^{s,t} = \mathrm{Ext}_{\mathcal{A}}^{s,t}(H^*(Y), H^*(X)) \Rightarrow [X, Y]_{t-s}$$
are displayed with the topological degree $t - s$ horizontally and the homological degree s vertically.

We are particularly concerned with the special case when Y is the Adams summand l of ku when the spectral sequence reads
$$E_2^{s,t} = \mathrm{Ext}_{E(1)}^{s,t}(\mathbb{F}_p, H^*(X)) \Rightarrow l^{s-t}(X),$$
and is module valued over the case $X = S^0$,
$$E_2^{s,t} = \mathbb{F}_p[a_0, u] \Rightarrow l^*$$
where $a_0 \in E_2^{1,1}, u \in E_2^{1,2p-1}$.

The rest of this section is restricted to the prime $p = 2$, where the spectral sequence reads
$$E_2^{s,t} = \mathrm{Ext}_{E(1)}^{s,t}(\mathbb{F}_2, H^*(X)) \Rightarrow ku^{s-t}(X)$$

and the corresponding spectral sequence for homology reads
$$E_2^{s,t} = \mathrm{Ext}_{E(1)}^{s,t}(H^*(X), \mathbb{F}_2) \Rightarrow ku_{t-s}(X)$$
(where the $\mathcal{A}$ action on $H^*(X)$ is twisted as detailed in Section 2.1 when determining Bockstein differentials). These are both module valued over the case $X = S^0$,
$$E_2^{s,t} = \mathbb{F}_p[h_0, v] \Rightarrow ku^*$$
where $h_0 \in E_2^{1,1}, v \in E_2^{1,3}$.

The conventions have the effect that for a space X, the non-zero entries for calculating $ku^*(X)$ are on, above and to the left of the line of slope $1/2$ through the origin. The Adams spectral sequence
$$E_2^{s,t} = \mathrm{Ext}_{E(1)}^{s,t}(H^*(X), \mathbb{F}_p) \Rightarrow ku_{t-s}(X)$$
is concentrated in the first quadrant at E_2, while at E_∞ it will be concentrated below and to the right of a line of slope $1/2$ with s intercept roughly the exponent of the p-torsion in the integral homology of X.

Each dot or open circle in the displays corresponds to a basis element of the corresponding $E_r^{s,t}$ as an $\mathbb{F}_2$-vector space.

Multiplication by h_0 is indicated by a vertical line, and a dotted vertical line indicates an 'exotic' extension not arising from multiplication in E_2.

Multiplication by v is indicated by a line of slope $1/2$ and is explicitly shown only at the bottom of an h_0-tower.

APPENDIX B

Indices.

B.1. Index of calculations.

We summarize where to find the calculations for various specific groups. The blanks in the first two columns correspond to well known facts we have not recorded. The blanks in the last two columns refer to calculations we have not done.

Group	Name	$R(G)$	$H^*(G)$	$ku^*(BG)$	$ku_*(BG)$	$t(ku)_G^*$
C_n	cyclic	2.2	2.2	2.2^4	3.4.3	3.6.2
$G_{p,q}$	non-abelian pq	2.3	2.3	$2.3.2^1$	-	-
Q_8	quaternion	2.4	2.4	2.4.6	3.4.3	3.6.2
Q_{2^n}	quaternion	2.4	2.4	2.4.5	3.4.3	3.6.2
$SL_2(3)$	special linear	-	-	2.4.12	-	-
$SL_2(\ell)$	special linear	-	-	$2.4.11^2$	-	-
D_8	dihedral	2.5	2.5	2.5.5	3.5.1	-
D_{2^n}	dihedral	2.5	2.5	2.5.4	-	-
A_4	alternating	2.6	2.6	2.6.2	-	-
Σ_q	symmetric, q prime	-	-	2.3.3	-	-
V	elementary abelian	-	4.2	4.2^3	4.1.1	4.13.1

Notes: (1) Special cases $p = q - 1, (q - 1)/2$ and $p = 2$ are given more explicitly in 2.3.3, 2.3.4 and following discussion.
(2) Only 2-local information.
(3) The ring structure is determined by 4.2.6, but no presentation is given.
(4) An alternative presentation is given in the paragraph preceding 2.3.3.

B.2. Index of symbols.

$\odot$		formal multiplication $x \odot y = x + y - vxy$
$[n](x)$		$(1 - (1 - vx)^n)/v$ the multiplicative n-series
$M^\vee$		vector space dual $\operatorname{Hom}_k(M, k)$
$\lvert V \rvert$		V with the trivial action
$W \uparrow^G$		induced representation
$[M]$		Hilbert series
$(1-x)^r_{[i]}$	4.8.1	truncation of $(1-x)^r$
$M(n)$		n-fold suspension of M
$\hat{x}$		natural representation of the quotient by $\langle x \rangle$
$X^{(n)}$		n-skeleton

117

B.3. Index of notation.

A

a	2.4	(1) generator of $ku^*(BQ_{2^n})$, $a = e_{ku}(\psi_0 - 1)$
	2.5	(2) generator of $ku^*(BD_{2^{n+2}})$, $a = e_{ku}(\hat{st})$
a_0		Adams spectral sequence counterpart of p
A_4		alternating group of degree 4
α		generic one dimensional representation
$\mathcal{A}$		mod p Steenrod algebra

B

b	2.4	(1) generator of $ku^*(BQ_{2^n})$, $b = e_{ku}(\chi)$
	2.5	(2) generator of $ku^*(BD_{2^{n+2}})$, $b = e_{ku}(\hat{s})$
B_i	2.2, 2.3, 4.2	indecomposable summand of BC_p
$BSS(a)$	1.4.1	Bockstein spectral sequence for a
$BP\langle n \rangle$		coefficient ring $\mathbb{Z}_{(p)}[v_1, \ldots, v_n]$
β		(1) generic one dimensional representation
	2.3	(2) p-dimensional representation of $G_{p,q}$

C

c_i^H	1.3	Chern class in $H^*(BG)$
c_i^R	1.3	Chern class in $R(G)$
$c_i^{K_G}$	1.3	Chern class in K_G^*
c_i^K	1.3	Chern class in $K^*(BG)$
c_i^{ku}	1.3	Chern class in $ku^*(BG)$
C_n		cyclic group of order n
$\hat{Ch}(G)$	1.1	the completed character approximation
χ	2.4	(1) representation of Q_{2^n}
	4.9	(2) Euler characteristic

D

d	2.5	generator of $ku^*(BD_{2^{n+2}})$, $d = e_{ku}(\sigma_1)$
d_i	2.5	element of $ku^*(BD_{2^{n+2}})$, $d_i = e_{ku}(\sigma_i)$
D_{ku_*}	3.3	$D_{ku_*}(\cdot) = \mathrm{Hom}_{ku_*}(\cdot, ku_*)$
D_{2n}		dihedral group of order $2n$

E

e_H	1.3	Euler class in $H^*(BG)$
e_R	1.3	Euler class in $R(G)$
e_{K_G}	1.3	Euler class in K_G^*
e_{ku}	1.3	Euler class in $ku^*(BG)$
e_K	1.3	Euler class in $K^*(BG)$

$E(t_1, \ldots, t_n)$		exterior algebra on generators $t_1, \ldots, t_n$
$E(1)$		exterior algebra $E(Q_0, Q_1)$
EG		contractible free G-space
$\widetilde{E}G$		the join $S^0 * EG$
ϵ		the trivial one dimensional representation
η		the generator of the stable 1-stem
$\mathcal{E}(G)$	2.1	ideal generated by Euler classes

F

$\mathbb{F}_p$		field with p elements

G

G		generic notation for a finite group
$G_{p,q}$		non-abelian group of order pq
$\mathbb{G}_m$	1.3	multiplicative group scheme
γ_t	1.3	Grothendieck γ operations
$\Gamma_I M$		I-power torsion in M

H

h_0		the Adams spectral sequence counterpart of 2
$H_I^*(M)$	3.1	local cohomology of M
$H\mathbb{F}_p$		mod p Eilenberg-MacLane spectrum
$H\mathbb{Z}$		integral Eilenberg-MacLane spectrum
$H\mathbb{Z}_p^\wedge$		p-adic Eilenberg-MacLane spectrum

I

i_V		the inclusion $S^0 \longrightarrow S^V$
I		augmentation ideal $I = \ker(ku^*(BG) \longrightarrow ku^*)$

J

J		augmentation ideal $J = \ker(R(G) \longrightarrow \mathbb{Z})$

K

k		(1) generic notation for a field
	1.1	(2) gencric notation for a ku_*-algebra
K		periodic complex K-theory
ku		complex connective K-theory
ko		real connective K-theory
κ	1.5	Künneth comparison map

L

l	2.1	Adams summand of ku
L	2.1	string module
L_A^m	[27]	universal ring for multiplicative A-equivariant formal
λ^i	1.3	group laws i-th exterior power

M

$MRees(G)$	1.3	modified Rees ring, generated by Chern classes

N

$nil(S)$	1.1	topologically nilpotent elements of S
$\nu_p(n)$		largest power of p dividing n

O

P

P	4.2	(1) the polynomial ring $\mathbb{F}_p[y_1, \ldots, y_r]$
	4.3-4.11	(2) the polynomial ring $\mathbb{F}_2[y_1, \ldots, y_r]$ where $y_i = x_i^2$
	3.5	(3) $R/(2,v)R$, where $R = ku^*BD_8$
	3.2, 4.12-4.13	(4) $\mathrm{Start}(1)(2^{r-1}H_I^1(Q))$
		(5) generic Sylow p-subgroup
PE	4.6	the polynomial ring $\mathbb{F}_2[x_1, \ldots, x_r]$
$P\mathbb{Z}$	4.4	the polynomial ring $\mathbb{Z}[y_1, \ldots, y_r]$
$\wp_C$	1.2	prime with support C
ψ_i	2.4	representation of Q_{2^n}

Q

q		(1) $q = 2(p-1)$
	2.4	(2) generator of $ku^*(BQ_{2^n})$, $q = e_{ku}(\psi_1)$
q_i	2.4	element of $ku^*(BQ_{2^n})$, $q_i = e_{ku}(\psi_i)$
Q		the image of $ku^*(BG) \longrightarrow K^*(BG)$, usually the completed modified Rees ring
Q'	4.4	part of Q in degrees below $-2r$
Q''	4.4	quotient of Q by Q'
Q_{2^n}		quaternion group of order 2^n
Q_0		Bockstein, Sq^1 if $p = 2$
Q_1		First higher Milnor Bockstein, $Sq^1Sq^2 + Sq^2Sq^1$ if $p = 2$
q_S	4.2	$Q_1Q_0(\prod_{i \in S} x_i)$

R

r		p-rank of group
R		generic notation for a ring, or for $ku^*(BG)$
$R(G)$		complex representation ring

B.3. INDEX OF NOTATION.

$Rees(R,J)$	1.3	Rees ring of R: subring of $R[v,v^{-1}]$ generated by R, v and $1/v \cdot J$
ρ		(1) the regular representation
	2.4	(2) generator of a quaternion group

S

S	3.2, 4.12-4.13	(1) $\Gamma_{(v,|G|)}ku_*BG$
	4.2-4.11	(2) subset of $\{1,\ldots,n\}$
$\hat{s}, \hat{st}$	2.5	representations of D_{2^n}
$\mathrm{spec}(R)$		prime spectrum of R
$\mathrm{spf}(R)$	1.1	$\mathrm{Hom}_{cts}(R,\cdot)$ formal spectrum of R
$\mathrm{Start}(n)$		suspension so as to have lowest entry in degree n
$S(V)$		unit sphere in V
S^V		one point compactification of V
Sk^{2n-1}	1.5	Skeletal filtration $Sk^{2n-1} = \ker(ku^0(BG) \longrightarrow ku^0(BG^{2n-1}))$
σ_i	1.3	(1) ith symmetric function
	2.5	(2) representation of D_{2^n}
Σ		suspension

T

$\hat{t}$	2.5	representation of D_{2^n}
T		v-torsion submodule of $ku^*(BG)$
T_i	4.6	summand of T in the elementary abelian case
$t(ku)_G^*$	3.6	ku-Tate cohomology
$t(ku)_*^G$	3.6	ku-Tate homology

U

u		$u = v^{p-1}$

V

v		the Bott element
V		(1) generic notation for elementary abelian group
		(2) generic notation for complex representation
V_4		Klein 4-group

W

W		generic notation for a complex representation

X

x_i		codegree 1 generator of $H^*(BV)$
$X(G)$	1.1	the formal spectrum of $ku^*(BG)$

Y

y		Euler class as generator of cohomology
y^*	4.4	modified reduction of J
y_i		Euler class of α_i
y_S^n	4.4	

Z

ζ		root of unity
$\mathbb{Z}$		the integers
$\mathbb{Z}/n$		the cyclic $\mathbb{Z}$-module of order n
$\mathbb{Z}_p^\wedge$		the p-adic integers

B.4. Index of terminology.

A

Adams spectral sequence	2.1
alternating group, A_4	2.6

B

Bockstein spectral sequence	1.4	
Bott element		v

C

Chebyshev polynomials	2.4.4, 2.5.3	
Chern class	1.3	characteristic class of a representation
codegree		cohomological degree, upper index
Cohen-Macaulay		ring with depth equal to dimension
connective		homotopy zero in negative degrees
connective K theory	0.1 et seq.	
cyclic group		C_n

D

degree		homological degree, lower index
dihedral group		D_{2n}
duality	3.3, 4.12	

E

elementary abelian		a group of the form $(C_p)^r$ for some p and r
Euler class	1.3	
exotic	4.2	unexpected to the naive

G

Gorenstein	4.3	duality condition for rings
– in codimension i	4.1	duality condition for rings

H

Hilbert series	4.8	also known as Poincare series
homotopy Gorenstein	3.3	duality condition for rings up to homotopy

K

Künneth theorem	1.5	

L

lightning flash	2.1	certain $E(1)$-module
local cohomology	3.1	Grothendieck's method of calculating cohomology with support

M

minimal primes	1.2	
modified Rees ring	1.3.8	Subring of K_G^* generated by $1, v$ and Chern classes

N

nonabelian pq	2.3	A non-abelian semidirect product $C_q \rtimes C_p$

O

P

periodic K-theory		periodic complex K-theory $K^*(\cdot)$
Poincare series		See Hilbert series

Q

quaternion group		Q_{2^n}

R

rank		size of largest elementary abelian subgroup
Rees ring	1.3.8	Subring of $R[v, v^{-1}]$ generated by $1, v, 1/v \cdot J$

S

string module	2.1	certain $E(1)$-module

T

Tate cohomology	3.6	combination of $ku^*(BG)$ and $ku_*(BG)$

U

universal coefficient theorem	3.2

V

V-isomorphism	1.1	map inducing isomorphism of varieties

W

weird	4.2	Discussion of Wall's 'somewhat weird structure'

Bibliography

[1] J. F. Adams, "Stable homotopy and generalised homology", Chicago Lecture Notes in Math., University of Chicago Press, 1974.

[2] M. F. Atiyah, "Characters and cohomology of finite groups", *IHES Pub. Math.* **9** (1961), 23–64.

[3] Anthony Bahri, Martin Bendersky, Donald M. Davis, and Peter B. Gilkey, "The complex bordism of groups with periodic cohomology", *Trans. Amer. Math. Soc.* **316** (1989), no. 2, 673–687.

[4] Egidio Barrera-Yanez and Peter B. Gilkey, "The eta invariant and the connective K-theory of the classifying space for cyclic 2 groups", *Ann. Global Anal. Geom.* **17** (1999), no. 3, 289–299.

[5] Dilip Bayen, "Real Connective K-theory of Finite groups", Thesis, Wayne State University, 1994.

[6] Dilip Bayen and Robert Bruner, "Real Connective K-theory and the Quaternion Group", *Trans. Amer. Math. Soc.* **348** (1996), 2201–2216.

[7] D. J. Benson and J. F. Carlson, "Products in negative cohomology", *J. Pure and Applied Algebra* **82** (1992), 107–129.

[8] Robert R. Bruner, "Ossa's Theorem and Adams covers", *Proc. Amer. Math. Soc.* **127** (1999), no. 8, 2443 – 2447.

[9] A. Capani, G. Niesi and L. Robbiano, CoCoA, a system for doing Computations in Commutative Algebra, Available via anonymous ftp from: `cocoa.dima.unige.it`.

[10] Harvey Cohn, "A Second Course in Number Theory", John Wiley and Sons, New York, 1962.

[11] M. Cole, J. P. C. Greenlees and I. Kriz, "Equivariant formal group laws", *Proc. London Math. Soc.* **81** (2000), 355–386.

[12] Donald M. Davis and Mark Mahowald, "The spectrum $(P \wedge bo)_{-\infty}$", *Math. Proc. Camb. Phil. Soc.* **96** (1984), 85–93.

[13] J. Duflot, "Depth and equivariant cohomology", *Comm. Math. Helv.* **56** (1981), 627–637.

[14] J. Duflot, "The associated primes of $H_G^*(X)$", *J. Pure and Applied Algebra* **30** (1983), 131–141.

[15] W. G. Dwyer, J. P. C. Greenlees and S. B. Iyengar, "Duality in algebra and topology" Preprint (2002) 39pp

[16] A. D. Elmendorf, I. Kriz, M. A. Mandell, and J. P. May, "Rings, Modules and Algebras in Stable Homotopy Theory", *Amer. Math. Soc. Surveys and Monographs* **47** (1996), American Mathematical Society.

[17] A. D. Elmendorf and J. P. May, "Algebras over equivariant sphere spectra", *J. Pure and Applied Algebra* **116** (1997), 139–149.

[18] Z. Fiedorowicz and S. Priddy, "Homology of Classical Groups over Finite Fields and their associated infinite loop spaces", *Lecture Notes in Mathematics* **674** (1978), Springer-Verlag.

[19] William Fulton and Serge Lang, "Riemann-Roch Algebra", *Grundlehren der mathematischen Wissenschaften* **277** (1985), Springer-Verlag.

[20] Paul G. Goerss, "The homology of homotopy inverse limits", *J. Pure and Applied Algebra* **111** (1996), 83–122.

[21] D. J. Green and I. J. Leary, "The spectrum of the Chern subring", *Comm. Math. Helv.* **73** (1998), 406–426.

[22] J. P. C. Greenlees, "K-homology of universal spaces and local cohomology of the representation ring", *Topology* **32** (1993), 295–308.

[23] J. P. C. Greenlees, "Augmentation ideals in equivariant cohomology rings", *Topology* **37** (1998), 1313–1323.

[24] J. P. C. Greenlees, "Tate cohomology in commutative algebra", *J. Pure and Applied Algebra* **94** (1994), 59–83.

[25] J. P. C. Greenlees, "Equivariant forms of connective K-theory", *Topology* **38** (1999), 1075–1092.

[26] J. P. C. Greenlees, "Tate cohomology in axiomatic stable homotopy theory", Proc. 1998 Barcelona Conference, ed. J.Aguadé, C.Broto and C.Casacuberta, Birkhäuser (2001) 149-176.

[27] J. P. C. Greenlees, "Multiplicative equivariant formal group laws", *J. Pure and Applied Algebra* **165** (2001) 183-200.

[28] J. P. C. Greenlees, "Equivariant connective K-theory for compact Lie groups." (Submitted for publication) 17pp

[29] J. P. C. Greenlees and G. Lyubeznik, "Rings with a local cohomology theorem and applications to cohomology rings of groups", *J. Pure and Applied Algebra* **149** (2000), 267–285.

[30] J. P. C. Greenlees and J. P. May, "Generalized Tate cohomology", *Memoirs of the Amer. Math. Soc.* **543** (1995), American Mathematical Society.

[31] J. P. C. Greenlees and J. P. May, "Equivariant stable homotopy", Handbook of Algebraic Topology, North Holland (1995), 277–323.

[32] J. P. C. Greenlees and J. P. May, "Localization and completion theorems for MU-module spectra", *Annals of Math.* **146** (1997), 509–544.

[33] J. P. C. Greenlees and N. P. Strickland, 'Varieties and local cohomology for chromatic group cohomology rings' *Topology* **38** (1999), 1093–1139.

[34] Theodore J. Rivlin, "Chebyshev Polynomials", 2nd ed., John Wiley and Sons, New York, 1990.

[35] Shin Hashimoto, "On the connective K-homology groups of the classifying spaces BZ/p^{r}", *Publ. Res. Inst. Math. Sci.* **19** (1983), no. 2, 765–771.

[36] M. J. Hopkins, N. J. Kuhn, and D. C. Ravenel, "Morava K-theories of classifying spaces and generalised characters for finite groups", Proceedings of 1990 Barcelona Conference on Algebraic Topology, *Lecture Notes in Math.* **1509** Springer-Verlag (1992), 186–209.

[37] M. J. Hopkins, N. J. Kuhn, and D. C. Ravenel, "Generalised group characters and complex oriented cohomology theories", *Amer. J. Math.* **13** (2000), 553–594.

[38] David Copeland Johnson and W. Stephen Wilson, "On a Theorem of Ossa", *Proc. Amer. Math. Soc.* **125** (1997), 3753–3755.

[39] Gregory Karpilovsky, "Unit Groups of Classical Rings", Oxford University Press, 1988.

[40] Edmund Landau, "Elementary Number Theory", trans. by Jacob E. Goodman, 2nd ed., Chelsea Publ. Co., New York, 1966.

[41] W. H. Lin, D. Davis, M. E. Mahowald and J. F. Adams, "Calculation of Lin's Ext Groups", *Math. Proc. Camb. Phil. Soc.* **87** (1980), 459–469.

[42] W. Bosma, J. Cannon and C. Playoust, The Magma algebra system I: The user language, *J. Symb. Comp.*, **24**, 3/4 (1997), 235–265.

[43] H. R. Margolis, "Eilenberg-MacLane Spectra", *Proc. Amer. Math. Soc.* **43** (1974), 409–415.

[44] H. Matsumura, "Commutative Ring Theory", *Cambridge Studies in Advanced Mathematics* **8**, Cambridge University Press, 1986.

[45] J. P. May, "Equivariant and non-equivariant module spectra", *J. Pure and Applied Algebra* **127** (1998), 83–97.

[46] J. P. May and R. J. Milgram, "The Bockstein and the Adams spectral sequences", *Proc. Amer. Math. Soc.* **83** (1981), no. 1, 128 –130.

[47] Stephen A. Mitchell and Stewart B. Priddy, "Symmetric product spectra and splittings of classifying spaces", *Amer. J. Math.* **106** (1984), no. 1, 219 – 232.

[48] E. Ossa, "Connective K-theory of elementary abelian groups", Proceedings of 1987 Osaka Conference on Transformation groups, *Lecture Notes in Math.* **1375** Springer-Verlag (1989), 269–275.

[49] J. Pakianathan, "Exponents and the cohomology of finite groups", *Proc. Amer. Math. Soc.* **128** (2000), 1893–1897.

[50] D. G. Quillen, "The spectrum of an equivariant cohomology ring I", *Annals of Math.* **94** (1971), 549–572.

[51] D. G. Quillen, "The spectrum of an equivariant cohomology ring II", *Annals of Math.* **94** (1971), 573–602.

[52] A. Robinson "A Künneth theorem for connective K theory." *J. London Math. Soc.* **17** (1978) 173-181

[53] G. B. Segal, "The representation ring of a compact Lie group", *IHES Pub. Math.* **34** (1968), 113–128.

[54] G. B. Segal, "K-homology theory and algebraic K theory", *Lecture Notes in Math.* **575**, Springer-Verlag (1977), 113–127.

[55] N. P. Strickland, "Formal groups and formal schemes", *Contemp. Math.* **239** American Mathematical Society (1999), 263–352,.

[56] N. P. Strickland, "The $BP\langle n \rangle$ cohomology of elementary abelian groups", *J. London Math. Soc.* **61** (2000), 93–109.

[57] C. T. C. Wall, "On the cohomology of certain groups", *Proc. Camb. Phil. Soc.* **57** (1961), 731–733.

Editorial Information

To be published in the *Memoirs*, a paper must be correct, new, nontrivial, and significant. Further, it must be well written and of interest to a substantial number of mathematicians. Piecemeal results, such as an inconclusive step toward an unproved major theorem or a minor variation on a known result, are in general not acceptable for publication. Papers appearing in *Memoirs* are generally longer than those appearing in *Transactions*, which shares the same editorial committee.

As of June 1, 2003, the backlog for this journal was approximately 3 volumes. This estimate is the result of dividing the number of manuscripts for this journal in the Providence office that have not yet gone to the printer on the above date by the average number of monographs per volume over the previous twelve months, reduced by the number of volumes published in four months (the time necessary for preparing a volume for the printer). (There are 6 volumes per year, each containing at least 4 numbers.)

A Consent to Publish and Copyright Agreement is required before a paper will be published in the *Memoirs*. After a paper is accepted for publication, the Providence office will send a Consent to Publish and Copyright Agreement to all authors of the paper. By submitting a paper to the *Memoirs*, authors certify that the results have not been submitted to nor are they under consideration for publication by another journal, conference proceedings, or similar publication.

Information for Authors

Memoirs are printed from camera copy fully prepared by the author. This means that the finished book will look exactly like the copy submitted.

The paper must contain a *descriptive title* and an *abstract* that summarizes the article in language suitable for workers in the general field (algebra, analysis, etc.). The *descriptive title* should be short, but informative; useless or vague phrases such as "some remarks about" or "concerning" should be avoided. The *abstract* should be at least one complete sentence, and at most 300 words. Included with the footnotes to the paper should be the 2000 *Mathematics Subject Classification* representing the primary and secondary subjects of the article. The classifications are accessible from www.ams.org/msc/. The list of classifications is also available in print starting with the 1999 annual index of *Mathematical Reviews*. The Mathematics Subject Classification footnote may be followed by a list of *key words and phrases* describing the subject matter of the article and taken from it. Journal abbreviations used in bibliographies are listed in the latest *Mathematical Reviews* annual index. The series abbreviations are also accessible from www.ams.org/publications/. To help in preparing and verifying references, the AMS offers MR Lookup, a Reference Tool for Linking, at www.ams.org/mrlookup/. When the manuscript is submitted, authors should supply the editor with electronic addresses if available. These will be printed after the postal address at the end of the article.

Electronically prepared manuscripts. The AMS encourages electronically prepared manuscripts, with a strong preference for $\mathcal{AMS}$-LaTeX. To this end, the Society has prepared $\mathcal{AMS}$-LaTeX author packages for each AMS publication. Author packages include instructions for preparing electronic manuscripts, the *AMS Author Handbook*, samples, and a style file that generates the particular design specifications of that publication series. Though $\mathcal{AMS}$-LaTeX is the highly preferred format of TeX, author packages are also available in $\mathcal{AMS}$-TeX.

Authors may retrieve an author package from e-MATH starting from `www.ams.org/tex/` or via FTP to `ftp.ams.org` (login as `anonymous`, enter username as password, and type `cd pub/author-info`). The *AMS Author Handbook* and the *Instruction Manual* are available in PDF format following the author packages link from `www.ams.org/tex/`. The author package can be obtained free of charge by sending email to `pub@ams.org` (Internet) or from the Publication Division, American Mathematical Society, 201 Charles St., Providence, RI 02904, USA. When requesting an author package, please specify $\mathcal{AMS}$-LaTeX or $\mathcal{AMS}$-TeX, Macintosh or IBM (3.5) format, and the publication in which your paper will appear. Please be sure to include your complete mailing address.

Sending electronic files. After acceptance, the source file(s) should be sent to the Providence office (this includes any TeX source file, any graphics files, and the DVI or PostScript file).

Before sending the source file, be sure you have proofread your paper carefully. The files you send must be the EXACT files used to generate the proof copy that was accepted for publication. For all publications, authors are required to send a printed copy of their paper, which exactly matches the copy approved for publication, along with any graphics that will appear in the paper.

TeX files may be submitted by email, FTP, or on diskette. The DVI file(s) and PostScript files should be submitted only by FTP or on diskette unless they are encoded properly to submit through email. (DVI files are binary and PostScript files tend to be very large.)

Electronically prepared manuscripts can be sent via email to `pub-submit@ams.org` (Internet). The subject line of the message should include the publication code to identify it as a Memoir. TeX source files, DVI files, and PostScript files can be transferred over the Internet by FTP to the Internet node `e-math.ams.org` (130.44.1.100).

Electronic graphics. Comprehensive instructions on preparing graphics are available at `www.ams.org/jourhtml/graphics.html`. A few of the major requirements are given here.

Submit files for graphics as EPS (Encapsulated PostScript) files. This includes graphics originated via a graphics application as well as scanned photographs or other computer-generated images. If this is not possible, TIFF files are acceptable as long as they can be opened in Adobe Photoshop or Illustrator. No matter what method was used to produce the graphic, it is necessary to provide a paper copy to the AMS.

Authors using graphics packages for the creation of electronic art should also avoid the use of any lines thinner than 0.5 points in width. Many graphics packages allow the user to specify a "hairline" for a very thin line. Hairlines often look acceptable when proofed on a typical laser printer. However, when produced on a high-resolution laser imagesetter, hairlines become nearly invisible and will be lost entirely in the final printing process.

Screens should be set to values between 15% and 85%. Screens which fall outside of this range are too light or too dark to print correctly. Variations of screens within a graphic should be no less than 10%.

Inquiries. Any inquiries concerning a paper that has been accepted for publication should be sent directly to the Electronic Prepress Department, American Mathematical Society, 201 Charles St., Providence, RI 02904, USA.

Editors

This journal is designed particularly for long research papers, normally at least 80 pages in length, and groups of cognate papers in pure and applied mathematics. Papers intended for publication in the *Memoirs* should be addressed to one of the following editors. In principle the Memoirs welcomes electronic submissions, and some of the editors, those whose names appear below with an asterisk (*), have indicated that they prefer them. However, editors reserve the right to request hard copies after papers have been submitted electronically. Authors are advised to make preliminary email inquiries to editors about whether they are likely to be able to handle submissions in a particular electronic form.

Algebraic geometry to DAN ABRAMOVICH, Department of Mathematics, Boston University, 111 Cummington St., Boston, MA 02215; email: `abramovic@bu.edu`

Algebraic topology and cohomology of groups to STEWART PRIDDY, Department of Mathematics, Northwestern University, 2033 Sheridan Road, Evanston, IL 60208-2730; email: `priddy@math.nwu.edu`

Combinatorics and Lie theory to SERGEY FOMIN, Department of Mathematics, University of Michigan, Ann Arbor, Michigan 48109-1109; email: `fomin@umich.edu`

Complex analysis and complex geometry to DUONG H. PHONG, Department of Mathematics, Columbia University, 2990 Broadway, New York, NY 10027-0029; email: `phong@math.columbia.edu`

*__Differential geometry and global analysis__ to LISA C. JEFFREY, Department of Mathematics, University of Toronto, 100 St. George St., Toronto, ON Canada M5S 3G3; email: `jeffrey@math.toronto.edu`

Dynamical systems and ergodic theory to ROBERT F. WILLIAMS, Department of Mathematics, University of Texas, Austin, Texas 78712-1082; email: `bob@math.utexas.edu`

*__Geometric analysis__ to TOBIAS COLDING, Courant Institute, New York University, 251 Mercer St., New York, NY 10012; email: `colding@cims.nyu.edu`

Harmonic analysis to ALEXANDER NAGEL, Department of Mathematics, University of Wisconsin, 480 Lincoln Drive, Madison, WI 53706-1313; email: `nagel@math.wisc.edu`

Harmonic analysis, representation theory, and Lie theory to ROBERT J. STANTON, Department of Mathematics, The Ohio State University, 231 West 18th Avenue, Columbus, OH 43210-1174; email: `stanton@math.ohio-state.edu`

Number theory to HAROLD G. DIAMOND, Department of Mathematics, University of Illinois, 1409 W. Green St., Urbana, IL 61801-2917; email: `diamond@math.uiuc.edu`

*__Ordinary differential equations, and applied mathematics__ to PETER W. BATES, Department of Mathematics, Michigan State University, East Lansing, MI 48824-1027; email: `peter@math.msu.edu`

*__Partial differential equations__ to PATRICIA E. BAUMAN, Department of Mathematics, Purdue University, West Lafayette, IN 47907-1395' email: `bauman@math.purdue.edu`

*__Probability and statistics__ to KRZYSZTOF BURDZY, Department of Mathematics, University of Washington, Box 354350, Seattle, Washington 98195-4350; email: `burdzy@math.washington.edu`

*__Real analysis and partial differential equations__ to DANIEL TATARU, Department of Mathematics, University of California, Berkeley, Berkeley, CA 94720; email: `tataru@math.berkeley.edu`

All other communications to the editors should be addressed to the Managing Editor, WILLIAM BECKNER, Department of Mathematics, University of Texas, Austin, TX 78712-1082; email: `beckner@math.utexas.edu`.

Titles in This Series

787 **Michael Cwikel, Per G. Nilsson, and Gideon Schechtman,** Interpolation of weighted Banach lattices/A characterization of relatively decomposable Banach lattices, 2003

786 **Arnd Scheel,** Radially symmetric patterns of reaction-diffusion systems, 2003

785 **R. R. Bruner and J. P. C. Greenlees,** The connective K-theory of finite groups, 2003

784 **Desmond Sheiham,** Invariants of boundary link cobordism, 2003

783 **Ethan Akin, Mike Hurley, and Judy A. Kennedy,** Dynamics of topologically generic homeomorphisms, 2003

782 **Masaaki Furusawa and Joseph A. Shalika,** On central critical values of the degree four L-functions for GSp(4): The Fundamental Lemma, 2003

781 **Marcin Bownik,** Anisotropic Hardy spaces and wavelets, 2003

780 **S. Marmi and D. Sauzin,** Quasianalytic monogenic solutions of a cohomological equation, 2003

779 **Hansjörg Geiges,** h-principles and flexibility in geometry, 2003

778 **David B. Massey,** Numerical control over complex analytic singularities, 2003

777 **Robert Lauter,** Pseudodifferential analysis on conformally compact spaces, 2003

776 **U. Haagerup, H. P. Rosenthal, and F. A. Sukochev,** Banach embedding properties of non-commutative L^p-spaces, 2003

775 **P. Lochak, J.-P. Marco, and D. Sauzin,** On the splitting of invariant manifolds in multidimensional near-integrable Hamiltonian systems, 2003

774 **Kai A. Behrend,** Derived ℓ-adic categories for algebraic stacks, 2003

773 **Robert M. Guralnick, Peter Müller, and Jan Saxl,** The rational function analogue of a question of Schur and exceptionality of permutation representations, 2003

772 **Katrina Barron,** The moduli space of $N = 1$ superspheres with tubes and the sewing operation, 2003

771 **Shigenori Matsumoto,** Affine flows on 3-manifolds, 2003

770 **W. N. Everitt and L. Markus,** Elliptic partial differential operators and symplectic algebra, 2003

769 **Jie Wu,** Homotopy theory of the suspensions of the projective plane, 2003

768 **R. Höpfner and E. Löcherbach,** Limit theorems for null recurrent Markov processes, 2003

767 **Po Hu,** S-modules in the category of schemes, 2003

766 **Su Gao and Alexander S. Kechris,** On the classification of Polish metric spaces up to isometry, 2003

765 **Robert Bieri and Ross Geoghegan,** Connectivity properties of group actions on non-positively curved spaces, 2003

764 **J. Spandaw,** Noether-Lefschetz problems for degeneracy loci, 2003

763 **Yasuyuki Kachi and Eiichi Sato,** Segre's reflexivity and an inductive characterization os hyperquadrics, 2002

762 **Leiba Rodman, Ilya M. Spitkovsky, and Hugo Woerdeman,** Abstract band method via factorization, positive and band extensions of multivariable almost periodic matrix functions, and spectral estimation, 2002

761 **Oliver Druet and Emmanuel Hebey,** The AB program in geometric analysis : Sharp Sobolev inequalities and related problems, 2002

760 **Markus Banagl,** Extending intersection homology type invariants to non-Witt spaces, 2002

759 **Donald M. Davis,** From representation theory to homotopy groups, 2002

758 **Alan Forrest, John Hunton, and Johannes Kellendonk,** Topological invariants for projection method patterns, 2002

TITLES IN THIS SERIES

757 **Douglas Bowman,** q-difference operators, orthogonal polynomials, and symmetric expansions, 2002
756 **José Ignacio Cogolludo-Agustín,** Topological invariants of the complement to arrangements of rational plane curves, 2002
755 **M. A. Mandell and J. P. May,** Equivariant orthogonal spectra and S-modules, 2002
754 **Edward L. Green, Idun Reiten, and Øyvind Solberg,** Dualities on generalized Koszul algebras, 2002
753 **Daniel Panazzolo,** Desingularization of nilpotent singularities in families of planar vector fields, 2002
752 **Linus Kramer,** Homogeneous spaces, Tits buildings, and isoparametric hypersurfaces, 2002
751 **Bruce Allison, Georgia Benkart, and Yun Gao,** Lie algebras graded by the root systems BC_r, $r \geq 2$, 2002
750 **Masaki Izumi and Hideki Kosaki,** Kac algebras arising from composition of subfactors: General theory and classification, 2002
749 **Nanhua Xi,** The based ring of two-sided cells of affine Weyl groups of type $\widetilde{A}_{n-1}$, 2002
748 **Jürgen Ritter and Alfred Weiss,** The lifted root number conjecture and Iwasawa theory, 2002
747 **Armand Borel, Robert Friedman, and John W. Morgan,** Almost commuting elements in compact Lie groups, 2002
746 **Peter Niemann,** Some generalized Kac-Moody algebras with known root multiplicities, 2002
745 **Mikhail A. Lifshits and Werner Linde,** Approximation and entropy numbers of Volterra operators with application to Brownian motion, 2002
744 **Roger Chalkley,** Basic global relative invariants for homogeneous linear differential equations, 2002
743 **Heng Sun,** Spectral decomposition of a covering of $GL(r)$: the Borel case, 2002
742 **J. E. Gilbert, Y. S. Han, J. A. Hogan, J. D. Lakey, D. Weiland, and G. Weiss,** Smooth molecular functions and singular integral operators, 2002
741 **Francisco Santos,** Triangulations of oriented matroids, 2002
740 **Rick Durrett,** Mutual invadability implies coexistence in spatial models, 2002
739 **Georgios K. Alexopoulos,** Sub-Laplacians with drift on Lie groups of polynomial volume growth, 2002
738 **Yasuro Gon,** Generalized Whittaker functions on $SU(2,2)$ with respect to the Siegel parabolic subgroup, 2002
737 **Arjen Doelman, Robert A. Gardner, and Tasso J. Kaper,** A stability index analysis of 1-D patterns of the Gray-Scott model, 2002
736 **Wojciech Chachólski and Jérôme Scherer,** Homotopy theory of diagrams, 2002
735 **Martina Brück, Xi Du, Joonsang Park, and Chuu-Lian Terng,** The submanifold geometries associated to Grassmannian systems, 2002
734 **Michel Van den Bergh,** Blowing up of non-commutative smooth surfaces, 2001
733 **Milé Krajčevski,** Tilings of the plane, hyperbolic groups and small cancellation conditions, 2001
732 **Jan O. Kleppe, Juan C. Migliore, Rosa Miró-Roig, Uwe Nagel, and Chris Peterson,** Gorenstein liaison, complete intersection liaison invariants and unobstructedness, 2001

For a complete list of titles in this series, visit the
AMS Bookstore at **www.ams.org/bookstore/**.